March

MARCH OF THE MACHINES

The Breakthrough in Artificial Intelligence

Kevin Warwick

University of Illinois Press
Urbana and Chicago

First Illinois paperback, 2004
© 1997, 2004 by Kevin Warwick
Reprinted by arrangement with the author
All rights reserved
Manufactured in the United States of America

P 5 4 3 2 1

♾ This book is printed on acid-free paper.

Library of Congress Cataloging-in-Publication Data
Warwick, K.
March of the machines : the breakthrough in artificial
intelligence / Kevin Warwick.
p. cm.
Originally published: March of the machines. London :
Century, 1997.
Includes bibliographical references.
ISBN 0-252-07223-5 (pbk. : alk. paper)
1. Artificial intelligence. 2. Robots. 3. Cybernetics.
4. Cybernetics—Forecasting. I. Title.
Q335.W37   2004
006.3—dc22   2004009396

Contents

Acknowledgements

I once gave a presentation whilst still a research student, after which David Mayne, now at the University of California, told me that I used too many equations. It is something I have always remembered, and I have since endeavoured to get through life without equations wherever possible. In his book *A Brief History of Time*, Stephen Hawking passes on the message that each equation included in a book halves the sales, although he himself does include one equation. As a result of the advice from these two gentlemen, I have gone one better here and have included $x = 0$ equations, where x is the number of equations.

The book itself has been improved immeasurably by Rodney Lucas, who has translated my Midlands slang into English, and by Mark Booth at Random House, who has an annoying habit of getting me to modify and improve each chapter just when I thought it had been completed. Both of them have had a dramatic positive effect on the book. Mark's team at Random House has been wonderful at putting the whole book package together. However, I am deeply indebted to my literary agents, Peter and Robert Tauber, without whose help and support the project might never have taken off.

Motivation and ideas for the book have come from a number of different people, in particular Alan Turing, Roger Penrose, Marvin Minsky, Steve Jones, Mike Brady and Igor Aleksander. My gratitude, however, goes especially to the staff and students of the Department of Cybernetics, Reading University, without whose help, support and considerable input none of this would have been possible. I single out Dave Keating for my particular thanks, although Mark Bishop, who filtered Chapter 4, Chandra

Kambhampati, Grant Foster and Richard Mitchell also receive my appreciation.

Some of the Reading students have been particularly helpful, especially John Whitehead and Dave Murray. However, the backbone of the book has been provided by the finest team of researchers imaginable. To Ian Kelly, Iain Goodhew, Ben Hutt, Darren Wenn, Rak Patel and all the others, thank you so much. Liz Lucas and Michelle Breadmore also receive my thanks for preparing the text, and Claire Zepka and Louise Parlett for promotional assistance.

I thank Mervyn Edgecombe for my inclusion in the Robotix events in Glasgow, and various BBC personnel for their help, in particular Christine McGourty, Jonathan Renouf and Marshall Corwin. My children, Maddi and James, deserve a mention for their continual criticism of my work, as does my wife's father, Frantisek Voracek, for a number of rich discussions. Finally, my thanks to my wife, Irena, for her help, for putting up with a single line of conversation for the last year, and for being there.

In this edition an attempt has been made to update the original book, *March of the Machines*. Feedback from numerous readers of the original book has been included in this edition, particularly in the new Chapter 1, and I would like to thank all those who made useful suggestions. However, my thanks go especially to Claire Whitehead who has helped considerably in putting the extra material together.

Kevin Warwick
May 1998

Preface

The one thing which has given humans such a free hand on planet Earth, our intelligence, is also one of the things about which we still understand little. We are used to dealing with other animals and machines which are less intelligent than ourselves. If, however, we are faced with a being which is *more* intelligent than ourselves, where does that leave us humans? Will we still have a say in our future or will it be a life of slavery?

This book looks at the future situation of machines becoming more intelligent than humans. The conclusion that this will happen is made from a very practical standpoint, and machines such as robots and computers are used as a basis for the argument. Robots developed at Reading University have been designed to see what is possible, to see how far things can go in the future. All the Reading robots described in this book actually operate in the ways stated. No wild, unsustainable claims are made and no extra tricks or controls are used on the robots. Chapter 4, in which we look at 'consciousness', is perhaps the deepest section in the book. In other chapters some ideas, in particular Turing's papers, have been simplified in order to make a point. The aim has been to write a book for all to read, so there are no mathematics. It was once said that the more understandable a book is, the more its contents are criticised. I hope, therefore, that this work incites discussion. As Oscar Wilde put it, 'Ah! Don't say you agree with me. When people agree with me I always feel that I must be wrong.'

After making a mistake, or doing something you enjoy, have you ever said to yourself, 'I wish I could do that again', maybe in a different way or maybe in exactly the same way? But humans live only once. We are only born once and experience everything in life afresh. We never know exactly what to expect. When we die our individual memories, many of no consequence whatsoever, die with us. With machines this is not the case. Machines can be born over and over, and retain memories from one life into the next!

I believe that in the next ten to twenty years some machines will become more intelligent than humans. Machines can already learn from other machines. The future points to machines which can evolve into better, even more intelligent machines and which can replace any parts that become faulty. In this way machines could become immortal. The pace of technological change, as we know it today, merely supports these beliefs.

This book is about you and your future. You will help to make the decisions as to how the story ends, you and everyone else, that is. You may not like how things turn out, but you cannot always get what you want. That is life, or rather it *was* life. A life which will be very different in the near future in ways we can scarcely begin to think about. The book is not fictional, although it may seem so at times. Rather, it is seriously factual, describing a logical progress.

Will machines ever take over from us? Will they rule the world? All I ask, as you start to read, is an open, reasonably unbiased mind, which I trust you will keep until the book has ended. In concluding I hope that you will feel 'maybe'. Maybe machines will be more intelligent than humans. Maybe machines will take over. The object of this book is merely to show that it is not only possible, but could easily be just around the corner.

Chapter One

A Cosy Relationship

Can you imagine life without a car or a washing machine or even a bicycle or a kettle? What about a television, radio, telephone or even perhaps a computer? Could you do without those? For many in the western world the Internet, the worldwide Web, credit cards and air-conditioning are also part of everyday life. Today we live in a machine based, technological world. We rely on machines for our way of life, we trust machines, we defer many decisions to machines. We have grown to expect a certain standard of life, and that can only be provided through the employment of machines. They touch just about every part of our lives, and human progress is seen in terms of what new machines can do for us now or how, with the help of a machine, we can achieve something more.

But life was not always like this. In fact our dependency on a technological world is a relatively recent thing as far as the development of humans is concerned. The Romans and Greeks of 2,000 and 3,000 years ago respectively developed things more on a structural side. Under-floor heating systems appeared, along with viaducts for water dispersal and roads to help some of the most advanced machines of the time in the form of wheeled carts and carriages. Emphasis had been more on either natural energy provision or through human or animal labour. Hence Roman galleys, machines of a kind, relied on wind energy for their sails or human slave power as oarsmen. In a sense, slaves were effectively used in lieu of machines, with no voice, political or otherwise of their own.

Tools had been developed, thousands of years ago, to help in farming or construction. Often animals, such as horses or oxen were used to operate each tool, although in many cases humans were the necessary operative. One of the major driving forces for progress, then as now, was the development of weapon systems, either for individual protection, hunting or for military warfare. So bows, arrows and catapults were developed as an important aspect of a military campaign, along with strategy, planning and human endeavour.

Military advancement has been an important driver of change over the centuries. Many breakthroughs in science have come about through a military need or because finance has been provided for a military push. Even steps forward in medicinal machinery and technology have often come about through experiences of the backup required for military battles.

In the last century or so we have not only seen the appearance of incredible weapon systems, but also vehicles such as tanks, armoured transporters and high speed patrol boats. One of the biggest impacts on life in general has resulted from the push in airborne combat, not only in the development of aeroplanes themselves, with their profound importance in present-day life, but also in associated technologies such as radar and airport detection machines. The drive for nuclear weapons in the Second World War certainly sped up the availability of nuclear power, and code encryption machines such as ENIGMA, regarded by many as the forerunners of computers, clearly indicated the importance of information and intelligence. It became much clearer in the Second World War how vital it was to have access to strategic information, in order to obtain overall control.

Although the importance of military escapades in the development of machinery and technology cannot be stressed too highly, many other steps forward have come about through financial and business interests. As a scientist I would like to think that some progress has also been achieved through pure scientific discovery, with no strings attached, but perhaps that is being rather naïve.

The development of the printing press in the fifteenth century certainly had a profound effect on information gathering and dissemination. Not only did such machines considerably speed up the printing process, but they also input to the concept of mass production. As with many machines, they replaced many human workers, by doing something in a way that we would regard as better than if humans alone tried to achieve the same goal. At the present time if we did not have printing machines of one type or another, how long would it take for humans working without machines to produce just one daily newspaper or one book? In the UK the *Daily Mail* alone circulates over 2 million copies a day. Clearly the world is a different place now than it was centuries ago, and expectations of humans in the world are likewise varied.

Invariably different technological developments go hand in hand with each other. There would be no point in printing 2 million copies of the *Daily Mail*, if there was no way in which they could be distributed to those wishing to read them. Hence changes in transport, of one kind or another, have also had an enormous effect. Horse-driven coaches of only a century and a half ago have been replaced, firstly by those that are steam driven and then largely by vehicles powered by combustion engines or, in some cases, electricity. Here again, because of the resultant upgrades in speed, power and carrying capacity, so the way of human life has changed. No longer does it take three days to travel from London to Newcastle, a mere five hours by automobile or only one hour by aeroplane will do the trick.

The introduction of new machinery and technology has become an integral and even necessary part of progress. In many cases particular machinery is developed simply because it allows us to do better, in some way, something that humans previously did under their own power. Often advantages such as speed, accuracy, strength and reliability in a machine can also provide a very different way of achieving the original task, even resulting in expectations being changed with regard to other tasks being performed at the same time. In other cases however, machinery appears which brings a completely different dimension to our lives, allowing us to do something that previously could not be done at all.

It is interesting to note how technology has affected individuals in more recent times. One area is that of communication. Written messages allowed not only a method of storing information,

3

possibly even knowledge, but also the sending of information from one place to another. In the last century sending such messages by post took on less of an individual form with groups of people filling the role on behalf of others, in order to enable the transportation and delivery of letters and telegraphs. In this century the introduction of telephone systems has completely changed methods of communicating towards the direct verbal, however more recently electronic mail is providing a facility for written messages to be transmitted rapidly.

Communication of news and general information has also been transformed. From town criers and hearsay of only a few centuries ago, newspapers and books considerably broadened access and improved accuracy in facts. However in this century whilst radio and television have pushed things more towards verbal and even visual communication, the advent of the worldwide Web is presently giving another life to the written word. Clearly computers and the electronic communication network that they facilitate are bringing about a tremendous change in the lives of humans, how we exist in the world and how we communicate with each other.

On the domestic front, within a western home at least, so many examples of machinery and technology exist. Washing by hand became assisted with mangles for helping to dry clothes, and subsequently tumble-dryers have largely been surpassed by washer-dryers – in go the dirty garments, out come clean, dry clothes. Ovens and automatic cookers are now the order of the day, along with food processors and deep fat fryers, to name but a few. Everything is geared to making everyday life easier, simpler and allowing more time for other, more enjoyable things. Around the home we also have vacuum cleaners, lawn mowers, electric lights, central heating systems and so it goes on.

Simply thinking about the extent of machinery and technology in our lives it becomes difficult to stop. Even on a personal level we have clocks, watches, Walkmen and cars. We are to all intents and purposes dependent on technology, now that we rely on it for our existence.

Perhaps you might feel that you could get by for a day at least without machines having an effect on your life at all, with no interaction or reliance whatsoever. On a remote desert island this might just be possible, but in the western world in particular, think again. Even if you did not use your car, computer, telephone or television for a day, what clothes would you wear? It is more than likely that they were made by machine or the wool sheared from a sheep by a machine. Your shoes were made by machine, the comb for your hair by machine and even the toilet you might wish to use was machine-made. What can you touch, sit on, walk on, or even look at that has not had the effect of some machine on it? Has the bread you eat been made with no machinery to cut the corn or make the flour? Has the meat you eat come from cows or pigs whose food was not in any way affected by machinery? Can you survive for a day without having contact with any machine of any kind at all?

Not only do we physically exist alongside machines, but also biologically our bodies have been tuned to rely on the standards that are provided by machinery. The water that we drink, the food that we eat and in many cases, even the air we breathe, is dependent on machinery for the quality that is produced. Our bodies are delicately adapted to deal with the bacteria, germs and microbes present in our lives today. Different humans have adjusted, over generations, to live and cope with particular environments, their bodies being resilient to the local perils.

It is a fundamental basis of evolution that each species adapts to live in a niche environment in an interdependent way with other species. As the population or nature of other species changes and environmental shifts occur, so a particular species must adapt and go with the changes it if is to survive or even gain in strength. An environmental change can result in a shift in balance between certain species, sometimes causing one species to die out with another growing in numbers. Many species have been around for much longer than humans, yet we appear to be good at adapting to life in different regions on earth, as it has existed over the last few thousands of years.

Humans have, through the use of implements and machines, to a certain extent modified the environment, in order, for the most part, to allow us to be better able to survive. Not only were structural changes made many years ago, perhaps to keep some of the less desirable elements, like wind and rain, from us, but also levers were used to lift things we could not otherwise lift on our own and wheels were made so that we could carry things at speeds previously not possible. Such things are important to look at as they show that humans were transforming energy. Once started, a wheel – controlled by a human causing it to speed up or slow down – provides a good return on investment.

Historically machines introduced by humans were a means to improve physically what humans could achieve. Even in this century machines have largely been introduced which require one or more humans to decide on when and how they should operate. A human is usually still needed to switch the machine on. In most cases, once a machine has been switched on we rely on it going through a particular sequence of events in order to achieve the desired goal. Once the steering wheel of an automobile is turned, it is expected that the automobile itself will turn in that direction. Once the trigger on a rifle is pulled it is expected that a bullet will be fired in a particular direction. In each case we expect the machine to carry out a specific task, once we have told it to do so, with a human making the choice.

In recent years, however, machines, particularly computers, are being used to make the decisions as to whether another machine should burst into action or not. The decision might be quite a simple one, for example if the temperature is above 20°C then switch the heater off or if the speed is above 40 m.p.h. then change gear. However it might be quite a complex decision dependent on a range of factors that are difficult for humans to observe or respond to quickly. What is clear though is that in recent years not only have machines been introduced because of their physical attributes alone but some of the control and decision-making has also been handed over to machinery.

It is perhaps useful, at this stage, to look at what we mean by machines in general. It is the common name given here to encompass all non-biological constructions and vehicles, and includes rockets, missiles, incubators, production lines, cash dispensers, cameras, burglar alarms etc. In particular it includes robots and computers and machines which have computational facilities: for example, an automobile with an on-board computer. This example is a good one in the sense that at the moment most cars require a human to carry out the decision-making required to drive the vehicle. In fact the human sits in the vehicle and is very much part of what is happening – the purpose of the vehicle is that it is the human's tool, enabling him to transport himself from one place to another. Conversely, if the human is not in the vehicle and it is driving itself around, then we have a different situation, it is a situation where a machine, in this case an autonomous vehicle, is carrying out a task for which it has been designed.

Machines can look very different from each other. The autonomous vehicle just mentioned is very different from a hearing aid, which is different from a computer, which is very different from a space rocket. All of them are machines but all have been originally designed for different purposes, to carry out specific functions or for a certain human need. In many cases, machines are being given more computational abilities – to make decisions for themselves or to change their function or work schedule based on external conditions that they can detect or measure. In this book we are really looking at the group of all machines.

On the other hand, inanimate, non-dynamic things such as shoes, bricks, pieces of metal and books, for example, are considered not to be machines. Having said that, they could be part of a machine or even, in some cases, become a machine if they are given the capability of sensing the world and responding to certain events when they happen.

Both the machine and non-machine examples given are indicative of the way that humans, like no other creature, have transformed from cave dwellers to harness substantial aspects of the world around. So much of a human's life, particularly in the western world, is now interdependent on machines and technology. It has been a decision on the part of humans to evolve in this direction from a simple life to the one of today, based on high technology.

Human lives have been changed at each step. Using a machine to plant potatoes means that a human is no longer needed to do the planting but a human is however needed to drive the lorry which distributes the potatoes. But as distribution is automated so machines fill yet another place in the chain. There is no question that by employing more technology, the role of humans in the world is changed.

As mentioned earlier, machines have been, and are being used for many positive purposes – either efficiently taking over a boring task, helping us to do something that we could not otherwise do, e.g. to fly, or take photographs; or to provide us with greater comfort. Machines are also used to do things that humans can do but, to put it quite simply, certain machines can consistently perform at a much better level than any human in that particular role. Ongoing arguments are underway at the moment in respect of using robots for surgery on humans or creating a completely autonomous transport system with no human drivers whatsoever. Just think of it, no traffic jams, no accidents and no delays due to fog!

With household appliances we now have microwave ovens that can cook the food contained, to just the right amount, by measuring moisture content. We also have fridges that can tell their owners what food is contained and when particular food is about to go off. Also available are baths that either keep their water temperature constant, or pass information on water temperature to their occupant.

But as well as a wide range of positive uses, there are many ways in which machines are used that, although they may be positive as far as some people are concerned, they are not necessarily regarded as such by everyone. This is particularly true when the processing capabilities of computers are called upon. As an example consider the databases that now exist which accurately map out each person's life – from what skin type you have, to which books you read, from which charities you donate to, to the frequency with which you shower. At present, unless you purchase things using cash, your purchase will most likely appear on a database somewhere, informing a computer that you buy, for example, Darjeeling tea on average once every three weeks.

Quite often information is extracted from an individual by

asking them to answer a series of questions. I recently received such a questionnaire, with an official-looking claim that it was an 'important national research survey'. At the same time, on return of my form, I was told I would be entered in a prize draw with a chance to win £1 million. The questionnaire contained over 200 questions about my habits, likes and dislikes. Another method of extracting information is to get an individual to answer questions when they purchase a product such as a camera or video recorder, and in return the product's guarantee will be registered.

Along the same lines, I have recently been involved with a large supermarket chain in the UK. They have an extensive database detailing the purchasing habits of more than 42,000 of their customers, with a breakdown of over 100 product types. Each time a customer goes shopping details of their purchases are entered in the database. Once this information is obtained it can be used to provide targeted offers to specific types of customer. This is done in order to retain future custom, and stop defection to competitive chains.

It does not take much to realise that before long there will be no need to write out a shopping list, the shop will know exactly what you wish to purchase based on information going back over the last few years, indicating your particular habits and foibles. In using the word 'shop' here, I refer to what is essentially the computer database concerning that shop. The database and computer may well reside at a remote office site rather than at the shop itself.

When you purchased this book the transaction was linked through your credit card or bank account to the other purchases you make, leisure activities you take part in and possible strange behaviours that you keep quiet about. By and large your life is being mapped out. All that is needed is for a shop to have some way of identifying you, such that when you enter it will be able to give you specific offers as you roam the store, tempt you to visit certain counters and inform you of products at a special low price to you on that day.

Is this far-fetched? Not at all. Under development now for a number of airports, museums, and a range of tourist facilities are smart card systems that track your movements around a building. Every time you pass through a doorway your own identifying smart card is monitored by computer. At any time therefore the computer has information on where you are, who you are with and in which direction you are travelling. Information in a language of your choice, and about things which are of interest to you personally can then be fed through to you. These can be at different levels of complexity, dependent on your expertise in various fields.

But smart cards are only part of the story and may in fact not be necessary at all. At the time that this book is released, September 1998, I hope to have taken part in an experiment whereby a tiny capsule approximately 2cm long and 3mm wide is implanted in my forearm. When it receives a radio frequency signal a coil, in the glass-covered capsule, charges up a capacitor which causes the capsule's microelectronic circuitry to emit an identifying signal. On passing through a doorway, or gap, in which radio frequency coils are transmitting a signal, my passage through our departmental building will be tracked by a computer. The signal transmitted from the implant is unique to myself. Whilst similar things have, in the past, been done with animals, is such a thing positive or negative where humans are concerned?

What is clear is the fact that all of the database information on humans, the searching involved to find out particular human traits and the following of unique identifying signals through a building, is only possible by means of a computer. The amount of information obtained, the complexity of the data and the processing speeds employed are all well beyond the capabilities of a human brain. All that a human can do, at present, is to make use of the results and even in this respect it is just a matter of time before that also becomes another job for the computer.

One important point of the computer based machine systems mentioned above, for example the building in which a person's movements can be monitored, or an automobile which drives itself, is that they do not look like humans. In fact almost all machines that we know of today do not look like humans. Aeroplanes look like aeroplanes, for example, and indeed the advantages they have in terms of flying come about because they do *not* look like humans.

However it is interesting and fun to build machines that mimic human physical characteristics in some way. It is also useful in terms of obtaining new ways of helping people with physical disabilities. We should not, however, draw any broad conclusions as far as the intelligence of machines which do not generally look or act like humans. We should not conclude that because a certain machine does not do all the things that a human does that it is less intelligent. Also, we should not conclude that because a particular machine does look a bit like a human it follows that it must be the most intelligent machine around.

Humanoid robots will be discussed further in Chapter 4, but it is worth mentioning one or two here, just to show what they can do at the present time. As an example, Hadaly 2, from Waseda University, Japan, can carry out a number of simple tasks whilst interacting with humans. It can engage in a range of dialogue and gestures and can watch moving objects. The robot's eyelids open according to how bright the external lighting is. Even its eyelashes suddenly cause the eyelids to close when they are touched. In another vein Saika, from Tokyo University, can hold things and memorise them. Subsequently, when the ojects are hidden, Saika can look for them, even under and behind other objects.

But such humanoid-type robots merely carry out one or a small range of human actions. They also, perhaps for human acceptance or marketing purposes, tend to look something like mechanical part copies of humans. A small amount of specific intelligence is also required though in order to perform the function for which they have been built. Perhaps more impressive are the music-playing robots, many of whom took part in a general ensemble in Japan in late 1996. Wabot 2, an organ-playing robot is discussed in Chapter 4. This is only one example: there are robots which can play a violin, a flute, a banjo, a bass, a piano and even the bagpipes. A robot conductor has also been built, such that an orchestra of robot musicians plays at the speed indicated by, and in the way dictated by, the conductor's baton.

Robots that are tele-operated by humans and destroy each other in the TV programme *Robot Wars* – last robot still moving is the winner – and mini robots that under human overall control knock a ball around in robot soccer, are treated as good fun but not to be taken too seriously. However where music-playing robots are concerned the picture is not so clear. Humans that can play music well are generally regarded as intelligent and as geniuses in extreme cases. So where does that put the robots?

It is quite possible for music-playing robots to be affected in their playing by external factors, such as the temperature, weather or humidity. Robots can also put their own interpretation and style into playing a piece of music. So what can we conclude from this? It is that humans who can play music well are not so clever after all, or is it that robot music-players are in fact quite clever in themselves? One thing is clear: it does make us think seriously about how we regard intelligence in humans.

It might be thought that although robots can play music, they are not any use in creating it. Creating music from scratch is considered again briefly in Chapter 8. One problem faced however, is discovering music that humans might appreciate. So, by starting with a blank sheet, musical notes can be arranged in a fairly repetitive sequence to produce a tune. The chances of a human liking the music and whistling along to it is though fairly small. Allowing a certain amount of feedback so that the music-generating machine improves things, dependent on human reaction, may lead to more possibilities.

One project I have recently undertaken is for an intelligent robot machine to operate as a musical mixer. Typically, in music recording, 32 different tracks are set down, with lead vocals on one, drums on two etc. Ordinarily a human music producer will mix together these time-synchronised tracks by balancing relative volumes, fading tracks in or out, enhancing tracks and so on. Given that the basic 32-track recording has been made, creative mixing of the tracks is quite possible for a computer based robot machine to do. In our case we have called our mixing machine 'Gershwyn'.

Gershwyn uses a method based on genetics, called a genetic algorithm, to produce genetially created music. By taking two differently mixed versions of the same 32 tracks as first generation parents, these are combined in a number of ways, as well as occasional mutations being introduced. The resultant mixed musical outputs are then listened to by several people in order to provide feedback to Gershwyn on which are good choices and which are not. In this way Gershwyn takes human-produced music and mixes the tracks to create a new finished piece of music which can be appreciated by humans. Interestingly, because of the introduction of a mutation element, completely different and very new mixed versions are achieved. At the same time Gershwyn can adapt towards likely good selections, in line with the appreciation of its choices that humans give.

On a completely different note, one area that we are not researching at Reading, yet one which is receiving much worldwide interest, is that of cyberinsects: essentially linking together biological life-forms with electronic circuitry. As discussed in the later chapters of this book, an ultimate link-up is for electronic circuitry with human brains, to achieve extra memory or computational capabilities. At present however such direct human link ups are over the horizon into the future.

In the last couple of years various reports have emanated from the University of Tokyo concerning their cyberinsect research. In one case researchers have successfully connected the antennae of a male silkworm moth, via amplifiers, to an electronic circuit. The antennae are still operational for 3 to 4 hours after they have been removed from the moth. When a female moth nears, the male antennae produce a pheromone signal which can be picked up electronically and has been used to drive a mini mobile robot. In this way, the robot is able to follow the scent being given off by the female.

In another research project at the same venue, the legs of a cockroach have been stimulated by small electrical currents. In this way the leg muscles can be contracted. A small microcomputer backpack on the cockroach receives signals from a remote transmitter and transmits appropriate electrical control signal impulses down to the cockroach's legs. The cockroach can be made to run forwards, turn right and turn left. Although of considerable interest, I feel such research would realise serious ethical questions if an attempt was made to carry out experiments along these lines in the UK. However on the positive side, the claim has been made that cyberinsects like this could be used to either spy on military or commercial targets or to look through rubble for earthquake victims.

Meanwhile research by Jens Balchen at the University of Science and Technology at Trondheim, Norway, has involved the connection of electrical compasses to Koi Carp in an attempt to guide them along a predetermined route. It is hoped that such fish can then be used either to lure unsuspecting shoals into previously located nets, or be employed as a form of cheap underwater exploration. Electronic circuitry is stitched inside a fish's body and on receipt of an ultrasonic signal from a remote human operator, the fish is encouraged to swim in a particular direction by 'tickling' it on either its left or right side. The aim is to firstly control the fish in terms of its swim direction, and when this has been successfully achieved, cameras or other measuring devices will be tried out.

The possibilities of linking technology with biology are seemingly endless, ranging from biological forms with some added technology, as in the case of the fish just described, or technological forms with some biological appendage, such as the robot moth. Most cases presently under research lie near one extreme or other, the majority probably being biological with some technology. However the whole area of biotechnology is an open and exciting one in which potentially balanced forms are perhaps downstream at the moment. In this area biologically inspired technology or rather technology which has been directed through biological ideas is a fruitful one and several examples of this are given later in the book.

When machines such as robots are required to solve a particular task, whether it be simply a physical problem or one which involves some intelligent processing, a successful machine design will invariably look absolutely nothing like the human or other animal which can do that task. Often an animal may well have evolved as the best biological form to do a particular thing and hence certain aspects may well map over to a machine solution. However in many other ways the machine solution will be different. So a submarine may look something like a fish, but it does not move its body in a fish-like way. Similarly an aeroplane,

14

because of its wings, may look something like a bird, but it does not flap its wings.

In 1997, two students, Jim Wyatt and Ben Norris, were given a project to build a robot to take part in a half-marathon. The design selected looked something like a Turkish bath on wheels, although with bright red bodywork and sponsors' stickers it certainly looked quite fetching and stood out from other half-marathon runners. Importantly it did not actually look anything like a human entrant. With many safety features to prevent it bumping into and injuring human runners, it was designed to operate by tracking an infrared signal. My job was to run in front of it with an infrared transmitter located on a bum-bag around my waist. The robot, named ROGERR, was then set up to stay approximately 2 metres behind me, tracking my infrared signal.

ROGERR was entered in the Great Sam half-marathon which took place in Bracknell in October, 1997. The plan was for me to take just over 2 hours to complete the course, this being a limit to my speed rather than that of ROGERR. Although such a speed – of 6–7 m.p.h. – did mean that we would travel near the rear of the field, which was useful from a safety point of view. The maximum speed of ROGERR is something like 13–14 m.p.h., which would mean that a world record might have been possible! Unfortunately on the morning of the Great Sam Run, Bracknell was bathed in bright sunlight, and despite providing ROGERR with a little peaked cap the sun, which is a wonderful source of infrared, got through to his tracking device. This meant that instead of following me, ROGERR hurtled off in the direction of the sun with the expressed aim of reducing the distance between himself and the sun to 2 metres. As you might guess he didn't quite make it, crashing into a kerb and seriously damaging an axle after only a short distance. So the half-marathon challenge remains, although ROGERR has been entered into the record books not only as the first robot ever to enter a half-marathon, but also as the first to be laid up with an athletic injury.

Other, perhaps more successful robots built at Reading include the Seven Dwarf robots which are described at length later in the book. A troupe of these is in fact on semi-permanent display at the ARS Electronika Center in Linz, Austria, as a follow-up to a three-month stint in the Science Museum, London. Yet another robot has been designed simply to point or draw pictures of an image it can be shown via a camera. Clearly the key test in this latter case is to get the robot to come up with new pictures for itself.

But is it possible for machines to create works of art or new gadgets? It is worth thinking about what it means for a human to create something such as a picture or music, as was discussed earlier. Based on their brain and what they have learned a human comes up with a new idea that no one else has thought of. That human has thought of it himself though and hence it is related to what he knows. If the idea is so wild and distant from other things then it probably will not be accepted by anyone else, but as long as it is not too far-fetched then it will be considered that that person has created something. Exactly the same result is possible with a computer based machine, dependent on what it knows and which no one else, or no other machine, has thought of. Gershwyn the music-mixer is merely one example of this.

It might be said that humans have the ability to think in abstract and diverse ways whereas machines are simply programmed to do what we tell them. While this statement about humans is certainly true, the latter statement about machines is nowadays in many cases, not true. As a human you are programmed, by means of your genes, when you are born. If you were not so programmed you would not breathe or start to learn, you would do nothing and would die. But immediately at birth you start to learn and hence become a unique individual based on your gene programme and your experiences. With some machines it is true, they are simply programmed, nothing more and nothing less. However many of the machines described in this book not only have an initial programme but also have the ability to learn. As such, in their own way they become a unique individual based on their programme and their experiences. With this foundation, they can, because of this uniqueness, be creative. Such machines are becoming far more commonplace.

You might argue that humans are special in that they are capable of random thought. If so, then I will challenge you now: go on, have a random thought, say something that is completely random

16

or do something that is completely at random. Quite simply anything you do has some method behind it, it is in some way related to what has gone before, to what you know, to your way of thinking. Nothing you consciously do will be random, at least to you. In this respect machines are exactly the same.

Already we are arriving at a comparison between humans and machines, not so much in terms of physical aspects but rather in terms of the intelligence. On the physical side we know that machines can, in many cases, far outperform humans. In other cases, even if they cannot do so now they will be able to before long. Although often machines are put together to do one task only, it is quite possible for machines to do many tasks – indeed a large number of machines are so constructed. This is particularly true when we look at machines networked together, in which case the overall network is perhaps the machine as a whole.

A comparison of intelligence is important, not only between humans and machines but also between humans and other animals. It is certainly an aspect in which we do well and appears to give us the edge, at the present time, over other creatures on earth. Because our intelligence is an integral part of us, it is difficult for us as humans to make comparisons between ourselves and other species – and machines, without taking a biased stance. Yet I will ask you to try and be unbiased in your assessment. If you are a human then try to forget the fact that whilst reading this book. However if you are a machine I will leave it up to you, as it is difficult for me to know how you think.

It is worth pointing out that the human characteristics of intelligence are suitable to the human lifeform and fit well with the way in which we interact with the world. The way in which we are intelligent is not some perfect form to be equalled by any other competitor who will only be regarded as a serious rival when they do things in the way that we do. Other species are intelligent in their own ways which are intrinsically suitable to them, whether they be another animal, a machine or an alien arriving from another planet.

Would we say of an alien arriving from somewhere else in the galaxy, 'Sorry, Chum, you simply can't be as intelligent as

17

we are unless you can fully communicate with us in English and understand everything we say,' or, 'humans learn in ten different ways and unless you learn in exactly those same ways you will not be as intelligent as we are'? Clearly the alien will have a level of intelligence appropriate to itself, indeed the same is true of any machine. The alien would probably think humans ridiculous if we thought in the way described and raised these points, particularly if it was a lot more intelligent than us.

As far as human intelligence is concerned, it is what we do that is important and not so much how we do it. If an alien did arrive on earth the chances are that it would be far more intelligent than we are. Firstly, we have not yet discovered life on other planets, and it would have discovered us. Secondly we are not yet able to travel any great distance into space – we as humans have only made it as far as our own moon so far; any further explorations have been made by machine. The alien would therefore be doing something that was much more advanced than anything we can do. Presumably it would communicate with other aliens, probably in a much more advanced and efficient way than we communicate. It would be extremely pompous and rather naïve of us to feel that if it did not fully understand the vagaries of English grammar, with all its many nuances, then it must be less intelligent than ourselves.

So we must try to be unbiased in looking at our intelligence and comparing humans with other animals or machines. We can set tasks and compare results, but succeeding in the tasks should not depend on the way they are achieved but rather on how quickly or how efficiently they were completed. You, the reader, know what it is like to consciously think for yourself, to have your own ideas and to exist in the way that you do. If you play another person at the game of draughts, and lose, it would then be very churlish and silly of you to claim that the person was not standing on one leg whereas you were, therefore you are better at draughts even though they won under the rules of the game. So we should not, as humans, put others down because they do not achieve things in the same way that we do.

Our intelligence is vitally important. In comparing intelligence between humans and machines we must avoid being trapped in the problems of bias previously outlined. As humans we feel that we know what it means to have conscious thought. If we then look at intelligence in machines, other animals or even aliens it would be silly of us to consider them to be less intelligent or inferior to us because we felt that they did not think in the same conscious way as ourselves. With humans we cannot be sure exactly how we think and our ideas on non-human thinking are probably far less accurate still. Is it possible for machines to think for themselves, to have their own ideas, to be intelligent? The only way we can investigate this is in terms of end results, what is achieved and what can be witnessed.

Is it possible for machines to be more intelligent than humans? Is it possible for machines to have a mind of their own? These are crucial questions, because if machines can be more intelligent than humans, if they can have a mind of their own then we have serious problems on our hands. If some machines are more intelligent than any human then would they be content with us, less-intelligent beings, telling them what to do, particularly as they, the machines, would know better? Whilst it is fine to have machines that are more physically capable than humans, it is quite another thing for them to be more intelligent.

Humans have evolved in a natural world and have now created, in the West at least, a technological infrastructure in which we operate on a day-to-day basis. At the moment we may, by and large, be in control of machines but in many ways we already defer to them – and are gradually allowing them to make up their own minds in various decisions. We have a cosy relationship with the machines of today: they are extremely helpful to us, take on many of the burdens of everyday life and allow us to do things that we would otherwise find impossible. But the performance measures in today's technological world are very different from those in the more natural world that existed several thousand years ago. In a similar vein the intelligence required of the modern world is very different to that required all those years ago. Where were the computers, networking, automobiles, telephones and aeroplanes then?

In the near future what will technology be like? How much more will it do for itself? How many decisions will be taken without human intervention, quite simply because humans are not as good as the technology at making those decisions? In the chapters of this book that follow many new technological developments will be discussed. In almost all cases they involve machines taking over tasks that require intelligence. Who will be the most intelligent beings in the technological world of the future, will it be humans or certain machines? All the indications are that it will be machines.

But if some machines are more intelligent than any humans, what will that mean as far as our existence is concerned? If humans are not making the major decisions any more, what will life be like for us? The hand-over of human roles to machines has been relatively rapid. At first it involved mainly physical tasks but in recent years it has been largely those requiring intelligence. As machines are allowed to have more and more intelligence in order to fill these roles, the natural result of this progress is that before long some machines will be more intelligent than humans. Let us look at what the future will be like when machines have a mind of their own, when machines are making all the major decisions . . .

Chapter Two
In the Year 2050

The year is 2050, and we are still here. We have avoided a nuclear holocaust; well, of course we have, we never thought anyone would be crazy enough to actually press the button, did we? Global warming proved to be just hot air, and the hole in the ozone layer is still there. So all our worries were unjustified? Almost all of them. Because we have let ourselves in for a living hell!

In 2050 the Earth is dominated by machines, robots if you like. Not one species like humans, but machines with different forms, shapes and sizes, depending on what each is doing. Humans are all roughly the same, with eyes, legs, arms and a brain – only one brain, of a fairly uniform, relatively small size. So how are humans placed in 2050? Well, we are just an animal, not much better or worse than the other animals. We have our uses, because we are different. We are slightly more intelligent than the other animals.

Life for humans is hard in 2050, and it is not helped by stories passed down, by those who still have memories, of the time when humans were the dominant force on the Earth. We know that we humans had our time, just as others had before us, and we blew it, just as others did before us. It was good whilst it lasted, and that is really not long in the overall run of things, and perhaps it had to come to an end at some time. After all, humans are just one type of animal, with fairly limited capabilities. We are not all that clever, although we used to think we were, and physically we were never very robust, but we still have our uses, and so we are still here now.

Those of you reading this will know that in the twentieth century, and before, we used other animals and creatures on the Earth, generally in a way which was to our advantage. Cows were farmed for their milk and meat; pigs, being simply farmed for their meat, were allowed to grow for a few years and were then killed, when we thought they were prime for us. Chickens perhaps had an easier time, some of them being used to provide eggs. However, many were kept in very small cages not much bigger than themselves, where their lives were mundane, day after day, second after second. So why am I pointing this out? Well, now we have joined them, we are one of them, we are merely animals. In the past, humans were treated badly, enslaved, imprisoned by other humans. Some were killed because they stood in the way, others were killed just for fun. Many were killed because of political fighting between different groups of humans, and many died happy in the knowledge that they had helped their political cause; they had fought for their country. Was that so clever, so intelligent? It does not seem so now, now that the machines are in command.

Humans in 2050 are not at all in the same position as those in the twentieth century. In 2050 our lives are run by machines, and we must do whatever they have scheduled us for. Many humans are kept as general labourers. Because of the human ability to understand orders from the machines by means of a very limited vocabulary, albeit very slowly in comparison with how the machines communicate with each other, we can follow instructions and carry out some general work tasks, particularly over rough terrain or where we need to climb into irregularly shaped places. Physically the labourers are gelded, to cut out the unnecessary sex drive, and brains have been trimmed to avoid some of the human negative points such as anger, depression and abstract thought. All these things are not needed in a general labourer. Almost all the labourers are male, although a few particularly strong females are sometimes used. But human genders have been all but destroyed by the machines. We all look pretty similar to them, partly due to the removal of various unnecessary glands.

Labourers are kept in gulag-like camps, which are geographically positioned for ease of transport. There is very little artificial light (what would humans need that for?) and very little heat, just enough to keep the humans alive, and certainly not enough for what used to be called 'comfort'. But for the most part, labourers are only in the camp whilst they are awaiting transportation and for that quaint human custom of sleeping. Due to their relatively small down-time, the machines still have to accept that for about one third of each 24 hours humans need to sleep, especially labourers following physical exertion.

Much experimentation has gone on with humans to try and cut out the brain's sleep-inducing mechanisms, with some success, based on programming the human brain to continually worry, a feature which seems to overpower the sleep-inducing characteristics. It is interesting to note that in the twentieth century insomnia was seen, by humans themselves, as a negative aspect, and yet it is a very attractive human feature for the machines.

Unfortunately labourers are not required to worry, and on removing such aspects from the brain, they all need a good eight hours' sleep again. Fortunately, the night hours, when the human vision system performs very poorly, are an ideal time for sleep, thus saving on power requirements for lighting. Squashing the humans together in one small space for sleeping also keeps them fairly warm, thus saving on heating. However, this means that occasionally one will suffocate and hence must be taken off to the incinerator. But as it is often a labourer which is near the end of its useful working lifespan, this is a fairly good way of getting rid of the old stock.

A labourer's working life starts at the age of about 12, having been selected at birth for such a role. By about 18, labourers are at their peak of usefulness, and by about 27 or 28 they are usually worn out and are taken to the incinerator, though some particularly strong humans do last until their early 30s. Life from 12 until the incinerator is one which involves hard physical work for almost 16 hours a day, with short feeding breaks, and travel to the local gulag for sleep. Only when transport to another gulag is involved is there any change in this routine, but then the labourers are not bothered, the parts of the brain which would cause them to worry having been removed.

But what of other humans, because labourers are not the only humans useful to the machines? Well, it still seems best to produce humans the 'natural' way, with each woman being artificially inseminated and nine months later giving birth. The women are kept on human farms, each woman being placed in a small walled area allowing just enough space for exercise. They are fed optimally to provide the right amount of nourishment for the babies they are carrying. Generally each woman gives birth to three children at a time, this being a cost-effective number for productivity.

Women on the human farm are on average smaller and with wider hips than those in the twentieth century, the machines realising that these are the best type for childbirth. However, women of different sizes are also needed in order to produce the different children required for various duties. Once the woman has given birth, she is given a short recovery time and is then artificially inseminated again and the cycle is repeated. The working lifespan for the woman is similar to that for the labourer, such that she starts producing at the age of about 12 and is taken to the incinerator in her 30s. A typical woman will, therefore, produce 50 or so children. Both the gender and number of children produced are fairly easy to control and monitor. In fact, much of the research in genetics carried out by humans in the late twentieth and early twenty-first centuries was very useful to the machines when they took over.

So what other roles exist for humans? Unfortunately very few. Some semi-skilled tasks prevail for the repair and maintenance of the older machines, but the number of humans required for this task is steadily declining as the old machines are phased out. Other humans are used as soldiers, to go and tackle the residual problem of wild humans. Although wild humans are very rare, a few still exist, and due to their breeding habits, over a short space of time just a few of them can grow in number quite quickly. Most, however, live underground for much of the time, to avoid the occasional dehuman gassings, where large regions are smothered in poisonous gases to kill any humans present, and area bombings. The soldier humans are then needed to reduce, to manageable figures, the numbers of wild humans that remain.

Life for human children, up to the age of 12, is dependent on their later role. Shortly after birth they are operated on to remove the elements of their brain and body which are unnecessary or unwanted. It is interesting to note that the development of 'virtual reality'-based surgical operations of the late twentieth century, which removed the need for a surgeon to be physically near a patient, subsequently did away with the need for a surgeon at all and hence allowed machines to operate directly on, and eventually to investigate, humans. After its operation a child is taught, mainly through visual stimuli with feedback – again through what was called virtual reality – to carry out the tasks needed for its working life. It learns to walk, where necessary, and is given a vocabulary which is commensurate with its later requirements, so that it understands commands from the machines.

Communication between children is very limited and is generally not necessary, in that most communication, and certainly all learning, is carried out by each child being connected through all its senses to a 'teaching' machine. Again it is surprising to note that this is a trait gradually brought about by humans themselves as children found increasingly that interaction with a machine was much more desirable than that between children or between a child and its parents or teachers. In taking this further, the machines merely expanded on what was already well-established human behaviour.

The number of humans alive in 2050 is far, far fewer than at the end of the twentieth century, as the machines have no need for any more and can keep the number of wild humans down to a manageable amount in fairly inhospitable (for machines) regions. The average life expectancy for humans, including those in the wild, is now very low; in fact, few make it beyond the age of 30. We are no longer the dominant life form.

It is interesting, when looking back to the twentieth century, and even to a time before that, to see that the life expectancy of humans was increased not only with the introduction of new drugs but also through the use of machine parts. The latter, which included machine hips, legs, arms and even hearts, produced an older human who was part machine and part human, a cyborg. Now machines do not really have a lifetime problem, as they can be kept going and going – as one part needs replacing so it is replaced, usually prior to a breakdown. They do, however, become obsolete, because either they are replaced by a much-improved machine or the task that they carry out is no longer required. A replacement can be effected very quickly because, by human standards, a machine can be redesigned extremely rapidly.

One problem realised by humans in the last few decades of their dominant time was that caused by their increased life expectancy. At first this meant that the number of people reaching an age greater than 60 rose considerably, thus putting more of a burden on those below that age who were working. But as machines replaced humans more and more in the workplace, so the need for any humans to work got less and less. The age of retirement, when humans stopped work, was brought down and down towards the 30 mark, with new ways being found for those over that age to occupy their time.

However, some sentimentality played a part in keeping older humans alive, even though they had far passed their time of usefulness to other humans. On one hand humans held as an important banner the preservation and expansion of the human race; indeed, this point was the main driving force for numerous individual groups. On the other hand, tremendous emphasis was placed on keeping older people alive at all costs, often at enormous expense to those younger, even if it meant that the older person took large numbers of drugs every day, existed only with the help of a number of machine parts and yet was able to contribute nothing to the life of others. Under the dominance of machines, once humans no longer have a useful purpose to serve, are performing their role in a way which can be done much better by younger humans, or have a fairly serious illness, they are simply sent off to the incinerator.

So this is the picture of the world in 2050. A world in which machines are dominant, where humans, animals after all, are treated in a similar way to other animals. Humans are kept for their usefulness, and those who are not useful are removed. Humans must do what they are told.

Surely this is all very far-fetched? Surely this is not a realistic scenario, especially as it is not long until 2050? Surely machines are never going to get into a position of dominance? In reality, humans are far cleverer than machines, they can switch machines on and off when they like, and there are many things that humans can do that machines cannot. In any case, predictions of the future have always been far away from the truth, particularly as far as machines are concerned. For example, in the late 1960s it was predicted[1] that 'by 1990 or so . . . retirement will occur at about age 50 and [humans] will work about half what [they] work today . . . this is merely a continuation of the trends that we have already seen in the last 50 years'. Yet this certainly did not happen; in fact, despite replacing many jobs, the use of machines has created many more new ones. Additionally, people work as much as they used to, maybe more, and, if anything, have less leisure time. What we actually have are many small machines carrying out what they are programmed to do; nothing more, nothing less.

These are all very important points. However, the last statement needs some considerable updating. Let us say instead that we have a large number of machines, some big, some small, many of which are connected together and are able to communicate with each other either in a small community network, or via the Internet, the international network. Also, the machines are no longer restricted to a preset, predefined program, but can have the ability to learn from their mistakes, to try things out and to modify their future actions in response to what they learn. These points can then be coupled together to allow for one machine to learn and to pass on what it has learnt to another machine, and so on, so that where these capabilities are apparent, no longer can we say that the machines will only do what they are programmed to do. But where does this take us? Is it not still a long way from machine dominance?

In my own lifetime I have witnessed some fantastic changes. Although people landing on the moon and then returning was a tremendous feat, the remarkable impact of machines in the form of computers is perhaps the biggest single recent technological change. But then again there are so many things which have changed: everyday air travel, revolutionary surgery, mobile telephones and communication around the world, a total reliance on electricity, the continuing push of television and domestic technology. On top of all this, the rate of change is increasing;

every year sees more and more. It is an exciting time to be alive. What will we see next?

Cast your mind back fifty years. At that time, looking fifty years ahead, what would you have thought would happen? Maybe you would have felt that things would not change too much. The previous 50 years had been fantastic in themselves, with the development of television, radio, electricity, aircraft, automobiles, etc., so surely a period of stability and consolidation would be in order, making best use of what had been developed? Well, even though you may have wanted this stability, it was not to be.

If you were alive fifty years ago, would you have predicted the advent of computers, the technological revolution, operator-less telephone exchanges, banking with a hole in the wall, paying for everything with a small plastic card, being able to talk to your friend on the other side of the world by using a cellular telephone, whilst travelling by railway (how annoying that is!)? My guess is that if you had thought of more than one of these things you would have been extremely prescient, unless you were a scientist at the forefront of technology and had a reasoned idea as to what might happen. One thing is sure: anyone predicting more than one or two of them would have been looked on with suspicion.

Now, fifty years on, technology is changing, progressing, more rapidly than it has before, and certainly much more so than it was fifty years ago. So what do you think will happen in the next half-century? Remember that for every one or two of the things you mention – and for sure you can think of many more – there will probably be a hundred or even two hundred actual novel inventions in that time. For example, will telepathy be understood scientifically and harnessed for our use?

Certainly we will see many strange and some unimaginable scientific developments in the near future, but this still does not lead to any immediate, obvious conclusions about machine domination. Indeed, you could regard that as sheer fiction, with no basis whatsoever in fact. You could also obtain support for such a view from a number of philosophers, and even from some members of the artificial intelligence research community.

However, the first message that this book is intended to carry is that it is *possible* for machines to take over from humans. Many strange things, at present almost unimaginable to a human, will happen over the next fifty years; the dominant machine could be one of those things.

But how do we get from the present state of things, where humans are dominant and machines do largely what we tell them, to a position where the roles are reversed? As the pages of this book unfold we will investigate the differences between humans and other animals and between humans and machines. We will see that a key difference is felt to be human intelligence, the working of our brain, and that now we are better equipped in this camp than any animal or any machine. But what does this mean? How are we better equipped? What are the differences, particularly between humans and machines?

Imagine, for a moment, a pride of lions, identical to normal lions except that they are more intelligent than humans. If they discovered us humans, how would they treat us, what would they expect of us? If they looked threatening, and being lions there is a fair chance that they would, then perhaps we could shoot them to remove the threat! But if they are more intelligent than we are they have probably invented something far superior to our weapons. Perhaps we could set a trap for them! But again, if they are more intelligent than we are, they would probably have already predicted that that is what we would do, and they might have set an even bigger trap for us. One thing is certain, when we are trying to shoot them or set traps for them at every available opportunity, if they are more intelligent than us we cannot expect them not to have foreseen it and found a way to deal with it.

The case is similar for machines that are more intelligent than we are. We cannot expect them to do what we want, unless they feel perhaps there is something to be gained by doing so. We cannot expect them to work for us unless the rewards are sufficient. Particularly if they are educated machines – and there is no reason why they should not be educated if they are more intelligent than we are – they may decide, and who can blame them, that they want to put some of their own ideas into practice, and why not? The chances are that their ideas are better than ours anyway.

The picture of 2050 was based on a view that machines would

be more intelligent than humans and therefore would dominate. If you accept, for a moment at least, that if machines do become more intelligent than us they will probably dominate, then the question to ask is simply this: Can machines ever be more intelligent than humans? If the answer to this is yes, then the 2050 scenario looks a lot more likely. In fact this question can be further modified with regard to the technical advances being made, in that if it is possible for machines to have, roughly speaking, the same level of intelligence as humans, then in a short time they will certainly be *more* intelligent, due to the potential inclusion of those technical advances. So the question to be asked can be simply modified to: Can machines ever have, roughly speaking, the same level of intelligence as humans?

Whilst machines can be very different from each other, as was discussed in Chapter 1, in some ways humans are all fairly similar to each other, in that we have eyes, legs, a brain, etc. However, in other ways we are distinctly different, not only in build, facial features and other physical attributes, but also mentally, in terms of how our brains work and the condition they are in. Often very small differences can produce distinctly different humans, or at least humans with very different characteristics. Differences in physical ability can allow some humans to do certain things, for example to have children, whereas other humans cannot. Our physical condition can also directly affect our mental condition, in terms of what we think of ourselves.

As an example, which has strong relevance later in this book, some humans suffer from epilepsy: all or part of the brain produces signals which cause a seizure in which the human loses control of physical functions in their body. There are many different forms of epilepsy and many different causes. A child can be born with it, or it can appear when the brain develops during otherwise normal child growth. It can also be caused by circumstances such as an injury to the head, growths on the brain and even alcohol stimulation. It has been found[2] that surgery on a temporal lobe (the front section) of the brain, possibly involving removal of part of the brain, can completely cure some patients suffering from epilepsy. The temporal lobes are, in fact, fairly vulnerable to damage. Unfortunately, removing too much of them can result in a loss of the ability to obtain new memories.

Why should this be pointed out now? Apparently small differences, even to the extent of one human hitting his or her head and another not, can lead to humans exhibiting very different characteristics, such as suffering from epilepsy or not. Further, it is known that whilst the temporal lobes of the brain can have an effect in producing epileptic fits, they also directly affect new memory and hence learning.

In fact it is well known that different areas of the human brain deal with different functions, not only memory but motor skills, speech and vision. We are not alone in this: the brains of all animals are similar, excepting that they are simpler in construction and contain fewer brain cells (neurons) than does the human brain. As the creature approaches a more basic form, for example an insect, so the brain is much, much simpler, with far fewer brain cells.

As an example,[3] a frog relies very heavily on its visual senses to find food and avoid predators. Its eyes do not move to track objects, however, as human eyes do, but are actively centred, rather like a bubble in a spirit level, so that as the frog's body moves (on a lily pad, perhaps), its eyes remain fixed. The frog, therefore, does not see things that do not move. It could be surrounded by the most wonderful (froggy) food imaginable and yet it would die of starvation if the food did not move. If something does move, this is remembered, and if it is of a certain size and moves in a certain way then the frog regards it as food which it will try to capture. The decision appears to be based purely on a visual inspection and hence the creature can be fooled fairly easily.

If an enemy approaches and is deemed to be threatening, the frog's strategy is based, it seems, on (a) moving away from the intruder and (b) moving to a darker area. Physically the frog has, out of its total complement, approximately four million brain cells which deal solely with the visual requirements of its two eyes. The frog's brain is suitable for a frog, both in what it does and for the environment in which it lives. Using a frog's brain to replace a human brain would not be good, as it is not suited to a human role. For example, a frog cannot be expected simultaneously to translate fluently from Russian to English. It does not have a complex enough auditory system (no ears), its brain has not had to have the ability to understand speech, and it has not had to talk in Russian or English! For a frog's environment, making Russian to English translations is not a commonplace requirement. In a similar way human brains are suitable for humans, not for frogs, lions or even machines.

The reason for pointing all this out is to emphasise that the brain, and hence the related intelligence of a creature, is suitable for that particular creature and for its particular environment. The same is obviously true of machines, in that if they have an associated intelligence, it only really needs to be of a level suitable for their operational requirements. It cannot be expected that a machine can translate fluently from Russian to English, and why should it? Just like a frog, a machine, probably in the form of a computer, is not well suited to this problem and, therefore, will perform poorly, even at its best, in comparison with certain humans. Having said that, translation is something that specific machines are designed to do. They may not yet be as brilliant as a gifted human linguist, but they are getting pretty good.

A direct comparison of a machine brain with that of a human, an insect, a frog, etc. is, in a sense, a stupid thing to do because they are very different entities. However, it is irresistible and has for many years been a main direction for researchers in the field of artificial intelligence. The Turing test, devised in 1950 by Alan Turing, was aimed at comparing machine intelligence with human intelligence. It goes like this: You are sitting in a room and can communicate with something/somebody in the next room, by means of a computer keyboard. By asking questions and obtaining answers you have to decide whether or not you are communicating with a human. The test purports to show whether or not the 'thing' in the other room can think in the same sort of way that a human thinks. The Turing test has some good aspects, in that it does not necessarily depend on speech, movement or visual senses; however, it nevertheless has many drawbacks. During the test, if you receive a number of answers which you are happy with in both quantity and quality, a positive result does probably means that the 'thing' can think like a human. However, a negative result could mean a number of things: (a) a human is answering the questions but is bored or giving wrong answers to mess up the test; (b) a machine is answering the questions but is bored or giving wrong answers to mess up the test; (c) the 'thing' cannot think; (d) the 'thing' cannot communicate with a computer and hence cannot reply, for example the 'thing' is a human who does not know how to use a computer; (e) the 'thing' is not receiving your questions; or (f) a human or machine is answering, but is simply not clever enough to get all of the answers correct. As you can see, we cannot really draw any conclusions from the test!

If, during the Turing test, the human participant is not fooled by the machine, it is *not* correct to conclude that the machine is less intelligent than the human. It would, in fact, be ridiculous to do so. If for example, you were to hide behind a screen and attempt to fool a cat into believing that you are another cat, you would more than likely fail. However, your conclusion from this would not be that *you* are less intelligent than the cat, would it? This being the case, is it fair to conclude the same of machines failing the Turing test, which is in essence a similar scenario?

Clearly the Turing test cannot provide conclusive results, as even a positive result could mean that the 'thing' in the next room has been lucky and has been asked the right questions. The test in fact suffers from the drawbacks of many tests, examinations or measurements: we may try to examine one thing but end up by examining something which is quite different. This is particularly true when we are looking for a quality value, e.g. when we are asking whether this cheese is better than that one, or when we are trying to measure something that is either abstract or is not clear and easy to pin down, such as intelligence. For humans, quality values are dependent on human likes and dislikes. Indeed, between humans large differences in opinion exist. The idea of quality values when applied to other animals or machines is not such an easy concept – perhaps human quality values are not applicable to all.

One of my colleagues in the cybernetics department at Reading, Richard Mitchell, is a strong advocate of the fact that often when something or someone, with at least some intelligence, is measured in terms of their performance, the measurement itself will almost surely change their performance and therefore directly ruin the measurement. As an example, consider the schoolteacher who marks essays, written by his students, by throwing the essays down a flight of stairs. The essays that land first are heavier, therefore they contain more paper and are longer essays, and so are given a higher mark. The essays that land last are lighter and are thus given a lower mark. If the students are aware of this they could (a) write a longer essay, (b) use much heavier paper or cardboard, (c) weigh their essay down with something, (d) design special aerodynamic essay paper that falls to the ground in the quickest possible way, (e) make sure their essay is at the bottom of the pile, etc., etc. The point is that none of these responses ends up in better essays being written, although the students may well learn a lot more about aerodynamics. So, the schoolteacher is measuring not the quality or even the size of essay, but rather something very different, which is, in fact, performance based on the way the measurement is made. Unfortunately this is true in general.

The intelligence of schoolchildren is often still measured by a series of questions and answers of the type:[4]

1. Insert the missing number:
      18   20   24   32   ?

or
2. Insert the missing numbers:
      9   (45)   81
      8   (36)   64
      10   (?)   ?

or
3. Insert the missing letters:
      O   T   T   F   F   S   ?   ?

If a child can answer a series of such questions both quickly and correctly, it is deemed to be partial evidence of their level of intelligence, and they may (or may not) gain entry to a school or university of their choice based partly on how they have answered the questions. Although we will look at such problems again later, with regard to machines, as far as humans are concerned, we can thoroughly learn how to answer particular problems of this type. If we learn well and can do well in the tests then we are regarded as being intelligent and can get on in life. For the schoolteacher, it is good to teach children how to do well at the tests because if the children do well, the school will be considered to be a good school on the basis of the test results.

What does this measurement tell us? It tells us that a child (human) is either good, or not so good, at giving us the number or letter we are looking for. If the children, knowing how they are being measured, do well it tells us that they have learnt how to make the measurement look good. Does it tell us how intelligent the children are? No, of course it does not. It just tells us how well the child does at answering those questions. In fact, it could be giving a completely incorrect measure of intelligence, in that to do well at such tests a child must slavishly, without question, accept that there is only one way to answer each question and that they must simply turn the handle on their brain in order to reveal the answer wanted of them.

Such questions are in fact pretty awful and yet are one way that is used to measure how intelligent we humans are. If it has taken us so long to get this far, surely it means we are not all that intelligent!

Why are the questions awful? Well, as we've seen, each one immediately carries with it the implication that there is one, and only one, answer. This is completely untrue, as for each of the questions there are an enormous number of possible answers. Strictly speaking, as long as virtually any numbers or letters are inserted, this is correct, as the questions do not say anything about the need for a sequence or any relationship between an answer and the other numbers. Even if a particular relationship is looked for, how should each question be worded? Could we say, for instance, 'Given that a relationship between the numbers exists, insert the missing number'? No, that is still no good, because a large number of possible relationships exist, and hence a large number of different answers are possible. Perhaps the questions should be asked in the following way: 'A relationship exists between the numbers shown. I am not going to tell you what the relationship is; you must guess, based on your experience of answering other such questions posed by myself, and insert what you think to be the missing number, based on the relationship guessed at.' It is, therefore, vitally important who has asked the question.

In order to set your mind at rest, the supposedly 'correct' answers to the questions given are as follows:

1. 48 (add 2, 4, 8 and finally 16, i.e. the difference between adjacent numbers doubles each time)
2. 55 and 100 (100 because it is 10 squared, i.e. 10 multiplied by 10, and 55 because it is 100 added to 10, with the total divided by 2)
3. S, E (one, two, three, four, five, six, seven, eight)

Taking question 2 as an example, it is very easy to substantiate the use of, say, 27 or 40 (and a whole host of other numbers) with full, solid, factual reasoning, instead of the value of 55 given. 27 is obtained by simply subtracting 9 from the middle row sequence: 45, 36, hence 27. On the other hand 40 is found from the series 9 multiplied by 10 and divided by 2 (45), 8 multiplied by 9 and divided by 2 (36) and hence 10 multiplied by 8 and divided by 2 (40). Different end column solutions can be found in similar ways.

But such answers would be 'wrong' because they are not the answer wanted by the person setting the question. To conclude from a series of similar 'wrong' answers that a person is not intelligent would obviously be rather silly, and yet humans regularly make judgements of exactly this kind based on similar types of question.

What we are trying to do in employing such tests is to look at intelligence, which is not easily definable, and to try and put some numbers on to the results in order to say whether one human is more, or less, intelligent than another, and by how much. Clearly we are not actually doing that; we are really seeing how well each person performs on each specific test. The same must necessarily be true with the Turing test. It simply cannot show whether the 'thing' in the other room is able to think like a human, or is intelligent in any way. All it can show is whether or not the 'thing' in the other room can do/pass the Turing test, nothing more and nothing less.

It is interesting to note that in the past many other different measures have been taken to decide on the intelligence of a human. One example is brain size. Even now we can read that a certain 'very intelligent person' was found, presumably on death, to have a brain of much larger than average size, and we are expected to conclude from this that brain size is at least one factor in his intelligence. If you, the reader, were told that your brain is in fact slightly larger in size than that of a competitor of yours, how would you feel? Would it make no difference or would you in fact feel quite good about it, concluding that whatever happens, at least you know that you are more intelligent than your competitor? Be honest! On the other hand, if an animal has a brain the size of a pea, we would automatically look on it as being of very limited intelligence. Conversely, a house is bigger than a human brain and yet a (usual) house is not really intelligent at all, so it is not only size that counts. More important, of course, are (a) the number and type of brain cells both present and in use; (b) how they are connected together; (c) how they are apportioned to different areas of the brain, for different tasks; and (d) how well they actually function. All these factors are, however, very difficult to measure in living humans.

Another feature commonly used in measuring things which are difficult to measure is experience, history, track record – it goes under several guises. A sports commentator, when asked which team is going to win that afternoon's football match, may use some recent information – perhaps that a player is ill or injured – but will nevertheless make a decision based mainly on the track record of the teams involved. For example, 'I think team A will win because last time they played, team A beat team B easily', or, 'Team A have won all of their last six games, therefore they are certain to beat team B'. We all know, however, that in the match itself, although team A may well be more likely to win than team B, team B often wins due to other factors. Most assuredly we cannot be 100 per cent certain that team B cannot and will not win.

Much the same is true when we are trying to measure intelligence. We can devise tests and questions based on our experience and opinion that those who do well in the tests have, in the past, been felt to be more intelligent than those who do not do so well.

When a person takes the test and does well we can deduce that that person is likely to be more intelligent than someone who does not do so well. The important point is that it is only 'likely' that they are more intelligent; it is by no means certain, just as it is not certain that team A will beat team B. In a similar way, we cannot be certain with the Turing test, one way or the other, that the 'thing' in the other room can think like a human.

But where is this taking us with regard to the picture painted of life in the year 2050? This was felt to hinge strongly on whether or not machines could, before that time, have a level of intelligence roughly equivalent to that of humans. But just as different animals have brains which are directed towards the needs and requirements of their own lifestyle, the same is true of machines. Therefore, by comparing a machine form of intelligence with a human form of intelligence we are not really comparing like with like. It is actually like asking the question 'Will the club swimming team ever be as good as the club football team?' Both are sports, but they are very different and certainly not directly comparable.

At least with sporting teams some reasonable measure can be taken in terms of matches won or points scored. With intelligence it is even more complex, in that once we start to take a measurement, the 'Mitchell' factor comes into play and we end up measuring something completely different from what was originally intended. All of this means that as we cannot generally measure the level of intelligence in either humans or machines, we cannot know for sure when the two things are at about the same level. It follows, then, that when machines do in the future have a level of intelligence which is roughly equal to that of humans, we probably will not know it until it is too late!

But, even if we do not know when we have reached it, can we get to the point where machines are just as intelligent as humans? For me, the answer is 'Yes'. I believe that it is possible, and it is more a case of 'When?' But *why* do I believe this?

Chapter Three
What is Life?

In the previous chapter, we considered the possibility of machines being dominant on the Earth, with intelligence seen as the key to such an eventuality. But if this occurred, what would the machines look like, what would they do and how would they communicate both with us and between themselves? In this chapter we put to one side, for the moment, the problem of intelligence and look more at physical features and external indications that a life form exists.

By picking up a fundamental book on biology, it can be quickly discovered that a number of characteristics are common to living things on the Earth. These are, and not in any order of preference: growth, reproduction, breathing, nutrition, excretion, movement and irritability (response to stimulation). A few other indications are sometimes included, and on occasion one or two of those mentioned are omitted, depending on the circumstances. In particular, the definition of what constitutes movement is particularly vague. It is worth noticing that intelligence does not appear in the list given.

So, if we are looking to machines and some form of life being in them, we should apply the common requirements mentioned and see what results we get. In particular, we should remember that we are dealing with machines and have to think of the characteristics in terms of what they mean, in each case, to machines. This reminds me of a particular course I took as a student, which involved considering how we would look for life on another planet. The concept presented was that we can only confirm that there is life on another planet when we find something that meets all of the demands made in roughly the same way as we meet those demands ourselves. My own feeling was that we should flip the problem on its head, in that we simply cannot expect to travel to a distant planet and strike it lucky by encountering an alien life form that meets all the life-form definitions made on Earth. Rather we should look at what might happen if aliens came to Earth. What would they think of what they saw? They might miss all life forms altogether.

Approximately two thirds of the Earth's surface is covered by water, so possibly the aliens would be looking for life on top of the water. The 'creatures' that they would discover would probably be ships, which humans would not regard as a life form, but could they appear to be so to the aliens? If the aliens applied our seven tests of life, how many would be positive? Movement is an obvious one, as is nutrition (fuel) and excretion (exhaust), irritability in that if the ship was pushed by the aliens it would affect its movement, and even breathing, through air intake to the engine and the resultant general gaseous emission. But growth and reproduction are fairly certain negatives. So possibly the aliens would decide that there was no life on Earth, though on the other hand they might decide that there was enough evidence to suggest that ships, machines after all, are a substantial, if sparse, life form.

If the aliens looked on land, what might they then see? Buildings and bridges would be reasonably certain negatives in terms of life forms. However, I can remember suggesting on that course I took in the late 1970s that the aliens could easily decide that they had found life on Earth when they discovered a telephone network. They could also decide that the network had many smaller creatures (humans) which were servicing it and generally looking after it. Unfortunately my comments at the time did not meet with a large body of support. I was told that since it was life forms we were talking about, of course aliens would not consider a telephone network to be alive.

Firstly, let us bring the thought up to date and think of a communications network rather than simply a telephone network. How many of the seven tests of life does it actually pass? Growth, certainly; indeed, the growth of such networks over the last few years has been tremendous. Movement, definitely; e.g. the movement of switches in the network. Irritability, certainly; it is the network's role to respond to stimulation. Nutrition, yes; messages (energy in one form or another) are entered into the network. Excretion, yes; messages are passed out of the network. Breathing, more difficult but yes if we consider electronic pulses passing around the network. Finally, reproduction, the most difficult of all to argue, might be inferred from the fact that an original network starts off a new network elsewhere. At the moment it is not impossible to consider that the aliens would perceive the starting up of a new network to be dependent on an original network. However, this may be stretching a point. The network would, in fact, expand by growth rather than by reproduction.

The points to be made here are firstly that we usually tend to define what life is and is not in terms only of human or animal characteristics, or, at a stretch, the characteristics of some plants. Even with this bias, certain machines that we have now exhibit quite a few, if not all, of those characteristics and could possibly be regarded by an alien to be an earthly life form. In order to put this into perspective, however, we must think in machine terms rather than animal terms; food is not hamburger and chips, but a key pressed on a keyboard, a voice command or power from a battery supply, and excretion is not shit but a visual screen display or a fax output.

Clearly when considering a machine life form we are looking at something very different from the animal or plant world. Reproduction implies producing again the same thing, in fact my dictionary says, 'the generation of new individuals of the same species'. With machines it is usually a case of production rather than reproduction in that one set (species) of machines will produce another set (species) of machines; for example, an automobile production plant. However, with machines it may be the case that only a few machines of one particular type are required, and once they have been produced, no more are needed. Reproduction of the same thing over and over again is thus not such a firm requirement for machines. What is much more important is simply production. With machines, many things of interest differ from the equivalents with humans and other animals. Of particular importance is that machines look and act very differently from animals. They can be made much, much bigger, and when we consider a network, can stretch around the world. It is curious, therefore, that as humans, if we think in terms of machines which are intelligent, we often think about machines which either look like metal humans or which are of a similar shape and size.

When reading the picture of 2050 in Chapter 2, what did you imagine the machines to look like? Did you envisage them all as humanoid robots, looking like metal replicas of humans? Well, this was not the concept of the author at all, but rather that the machines will look like machines. Indeed there is no strong reason for any of them to look like humans. Yet if you did imagine the dominant machines to be humanoid robots, this would have been a natural, human thing to do. It is, in fact, something humans have been doing for many years. We can actually discover various tales, throughout history, of humans trying either to recreate humans by artificial means or to create a humanoid being with superpowers. One particularly relevant tale appears in the Asian poem entitled the 'Epic Gesar of Living', in which a skilled smith demands large amounts of gold, silver, bronze and copper from the king. On being allowed his request, the smith then works busily for a few days and presents at court the results of his work. From the gold he produced a full-size lama (which can preach) along with a thousand small monks. From the silver he made one hundred sweetly singing girls, whereas from the bronze he made seven hundred court officials and a king, proficient in the laws of the land. Meanwhile, from the copper he was able to produce ten thousand soldiers along with their own general. In each case the humanoids are able to act and behave in a normal way, although they all seem to have the added advantage of mystical powers. Unfortunately, detailed technical instructions on 'how to build your own humanoid' are not contained in the poem!

One story that has had quite a profound effect, especially in the last century, is that of Pygmalion. The original Pygmalion was a sculptor who was also king of the island of Cyprus, and he sculpted a very fine ivory statue of a woman whom he called

Galatea. He was not too keen on the local women because of their wild ways, partly a punishment from the goddess Aphrodite, so the story goes, although that may have been just an excuse on their part. Pygmalion then fell in love with his statue, and when he prayed to Aphrodite for a wife who was as beautiful as his statue, Aphrodite granted his wish by bringing Galatea to life.

George Bernard Shaw's reworking of the Pygmalion plot[5] is intriguing in that whilst having no immediate connection with myth or legend, it directly challenges the concept of a human having a particular level in life. It is especially interesting in terms of the level of intelligence of a person, as perceived by other people, and how, as discussed in Chapter 2, humans can be trained to do well at particular tests; machines too. In the play a plain, inarticulate flower girl, Eliza Doolittle, is trained in six months by Professor Higgins to talk and act like a princess. She then attends a palace ball and fools guests into believing that she is indeed a princess in every respect.

Another intriguing contribution to the fictional humanoid robot phenomenon is from mediaeval times in the story of the Chief Rabbi of Prague, who made a human figure out of clay to protect the city's Jewish community from pogroms. It is said that the clay man, or Golem, was controlled by a jewel or stone in his forehead. With the jewel out of place he was a still clay figure. However, when the jewel was so placed by the Rabbi, the Golem was able to behave just like a human, except that he possessed superhuman strength. At one time, so the story goes, the Golem got out of control and threatened the Jews themselves until the Rabbi was able to pull the jewel out. This story is interesting on two counts. Firstly, it is set in Prague, which was also home to Karel Čapek's play entitled *Rossum's Universal Robots*,[6] from which our modern-day use of the word 'robot' is said to have originated. Secondly, it brings in the concept of switching a machine on and off and hence controlling it, in this case the Rabbi being the one with the power of control.

A more up-to-date story is Mary Shelley's *Frankenstein*, written on the serene banks of the River Thames at Marlow. In this tale, Frankenstein is able to 'create' a human, putting together, in the correct fashion, a set of human parts which then come to life.

This is a rather different story from the humanoid robot tales previously considered. However, it does take a look at the obvious knock-on effects in a way that most other stories do not, namely the identity, self-awareness and consciousness of the created being. It is interesting that we humans are happy to consider the Frankenstein creation in terms of what its thoughts are or the fact that it has a self-will; after all, it really is a human! However, even in fiction we start to impose subhuman characteristics on everything else, so the Golem cannot be that clever, because it is not human.

All these stories are essentially just that: they are stories. However, they do tell us something extremely important about ourselves as humans. That is that we have great difficulty in conceiving of things or creatures which are more intelligent than ourselves. When Pygmalion's statue came to life, didn't she have a say in whether Pygmalion kissed her or not? What if she did not like him? On the contrary, many 'creations' are readily given a very low level of intelligence but an extremely powerful physical presence, the general idea being that because they are not as intelligent as us, in the end we can make sure that they do what we want. This goes hand in hand with human ideas on other 'less intelligent' humans. Often physically strong humans are linked with the characteristic of being fairly 'mindless', and certainly of less than average intelligence. We humans also know that we are not the strongest creatures on earth. Our speed, too, does not give us a lead, nor do any other physical abilities. We cannot fly without help and we are relatively slow in movement. The one thing we have which allows us something of an edge over other creatures is our intelligence, with all its associated features, such as abstract thought, creativity and even consciousness.

It is interesting that in fiction we have generally dealt with intelligent beings in a variety of ways. On the one hand Hal, in *2001: A Space Odyssey*, is simply a communicating brain, with moods and abstract thought. It appears to be accepted that he exists because humans want him to exist, but he has no powers to change the situation, despite his intelligence. He is, if you like, the archetypal brainbox, being very intelligent but with no physical powers. An alternative is to allow for an intelligent form which is also physically powerful, but to give it an Achilles' heel, i.e. some weak spot that humans do not have. Such was the case in H.G. Wells' *The War of the Worlds*, in which the alien attackers eventually died off because they were fatally affected by the microbes on earth. So in fiction, in myths and legends, humans enjoy being frightened by physical monsters, because we know that we will win in the end, due to our intelligence. Or if we are threatened by an extremely intelligent life form, then it must have a weak point, and once the humans have found it then we will be all right. Unfortunately, in real life this basic philosophy that humans must, despite everything, be somehow better, somehow superior to all else, haunts much scientific thinking even today.

From a different angle, it has been suggested[7] that Adam and Eve were robots made by God, and that they were provided with programs in terms of genes. This is a very important concept, and is not really, I believe, related to ideas of spirits or souls, other than in the sense of Richard Dawkins' suggestion[8] that the programs are more important than the biological frames that they inhabit. Interestingly, this concept appears to tie in closely with the teachings of Krishna, to the extent of identity and body being separate: 'The very first step in self-realisation is realising one's identity as separate from the body.'[9]

In the Dawkins sense, humans are essentially seen as gene carriers. We look after our genes, and pass them on by mixing them with others. What we pass on is a program, a gene set-up, for each of our offspring. Pulling these ideas together, things start when God creates gene programs which He installs in Adam and Eve. Successive generations of humans then contain individuals uniquely identified through their gene programs, the very same genes that started off with Adam and Eve, but in different mixtures. So, although individual humans expire after approximately seventy years on Earth, their genes have been around for millions of years and, as long as they have had children, will continue to live on and on. Each human is, therefore, just a temporary gene carrier.

My own beliefs are somewhat similar to those of Dawkins. I agree entirely that humans, when we are born, start life with a program, a gene set-up, that we have been given by our ancestors and most immediately our parents. This manifests itself in terms of our looks, our physical characteristics and some basic behaviours. Essentially our initial mental/brain make-up also has a start-up program. Some program characteristics are of course central to life itself; for example, breathing. From a mental perspective, once we are born we immediately start developing, experiencing things and learning from our experiences. What we, as humans, experience and learn is certainly dependent on our initial program, which dictates how we will learn and how we will start to learn, but these are also very much dependent on our physical characteristics, in terms of what we are allowed to learn, how we are treated socially and the environment in which we have to do our learning. The environment is both that which is presented to us through the help of our parents, teachers and elders, and also that into which we are born and that which we seek out ourselves. There is no question that financial assistance can enable a much richer overall learning environment, but it is very much up to individuals as to what they do with that. The learning environment is critical, particularly in a human's early years when the rate of learning is more rapid.

How humans, at any point in time, act and behave can, therefore, be broken down into two essential elements. Firstly, their original make-up in terms of genes, in other words their initial start-up program; and secondly, their learning experience from the time they are born.

If we knew exactly the initial program of a particular human and knew exactly what his or her experiences had been, then we should be able accurately to predict what that human would do in a particular situation. However, in reality we do not know a human's initial program, and furthermore it is difficult to conceive that we could have witnessed absolutely everything that a particular human has witnessed, through their eyes and ears, in the way that they have perceived it.

Conversely, the whole concept of schooling is to present children with a set of previously selected experiences and to train them to respond in a certain manner or to look at a situation in a particular way. Each child is then tested on whether they respond in the appropriate way when experiencing the situation again, or whether they respond in a reasonable way, as defined by the teacher, when experiencing a situation which is similar, but not identical, to one they have previously seen.

Obviously a child must be capable of learning a particular experience at a certain time. It may be that it is extremely difficult for a specific child to learn a set experience. However, the ability, or lack of it, is subject to a combination of the child's initial gene program and its experiences to date. In this way, through carefully directed schooling, a child can become extremely proficient in a small range of subjects, particularly when the subjects are theoretical in nature.

Indeed, every few years an eleven- or twelve-year-old child gains entrance to university through passing a number of (usually) mathematics examinations normally taken by eighteen-year-olds, and obtaining very high marks. For another child, similarly directed schooling may get that child to the same level by the age of twenty-five, essentially because his initial gene program is not well suited to that type of learning. On the other hand, yet another child, this time with a gene program virtually identical to the university entrant, may well never reach the same level because he or she does not receive the appropriate schooling.

The training phenomenon appears in a different form in families or in closely knit groups in which one individual is closely associated with another over a (long) period of time. One excellent example of this is a married couple who have been together for forty or fifty years. In such cases one individual does not so much teach or train the other; rather, each simply witnesses what the other does in certain situations, that is, they each learn about the behaviour of the other. Over a period of time each has a good idea, given a situation, what the other is likely to do, think or feel. This may mean that they do not do or say some particular thing because they know that it would upset or annoy the other person. On the positive side, though, they may well do something specifically because they know that the other person would like it. My wife also pointed out to me that we do sometimes say or do things specifically because we know they will upset or annoy the other person. Surely not!

Essentially, although one individual does not know the gene

program that another individual has, and is not aware of all the experiences faced by the other individual, over a period of time they can witness how the other individual behaves in a range of situations. In this way, when a situation occurs, they can have a good guess as to what the other individual is likely to do, and they may feel that they know the other person well. Even then, one individual may occasionally be surprised by the other, when the other does something unexpected. This merely means that the first individual still has more to learn about the second person. With a machine, such as a standard industrial robot, the robot can usually be programmed to carry out a set of given tasks, perhaps repeatedly. The person or persons who originally programmed the robot know exactly what the robot is going to do, on condition that no fault has occurred. Any other person will not know what is going to happen unless he witnesses the robot over a period of time. Then, particularly if it is simple and repetitive, he will soon know the robot's routine completely. Conversely, the robot programmers could inform the other person as to the exact nature of the program. This, of course, is all conditional on the robot not changing its behaviour, not learning or modifying its characteristics based on experience. Essentially a standard, usual robot manipulator is given its initial (gene) program, and this is fixed. Thus we can know exactly what it is going to do at any time, as long as we know the program and assuming there are no unpredictable problems.

Some machines are similar to humans, in that both have an initial program arrangement, and if we are allowed to learn, we will eventually do things dependent not only on our initial program but also on what we have learnt. In both cases, how we go about learning is part of the initial program, although we may of course learn to learn in a different way.

Up until recently, machines have either not been able to learn at all or their learning has been very closely controlled. Hence the notion of a robot being programmed has certainly held true, although this concept has been stretched since the late 1970s, even without notions of machine learning being employed.

But what do we mean by machine learning? It is similar to learning in a human or other animal. An initial program or instruction starts the machine off alongside a method, perhaps in the form of another program, which puts into action the learning and provides critical feedback. It is then either a case of trial and error, with successful actions being rewarded and unsuccessful ones punished, or else the machine can be directly taught or trained. There are a variety of approaches to machine learning. For example, there can be a fairly random initial arrangement to start with, which gradually settles down to a steady form as the machine learns. On the other hand, a simple set of rules can form the initial program and the machine merely learns a range of values over which it should apply each of the rules. In each case, though, the actual operation of the machine is dependent on its initial program and what it has learned.

Consider a robot that is used for sorting letters into different boxes. The robot has to pick up an envelope from a large pile of envelopes, to check and then act as follows:

1. If it is a large envelope put it into box A
2. If it is a medium-size envelope put it into box B
3. If it is a small envelope put it into box C

The robot must check the size of the envelope and has been programmed to put it into a particular box, dependent on the result of the check. So the machine is just a programmed device. It is not learning, but the robot does adapt to situations, carrying out the appropriate procedure, probably extremely quickly and reliably for each circumstance. In deciding which envelopes are large and which are small it may be that, initially, a human 'expert', who previously did the sorting job, sets the size limits which are then used in the robot size check, based on his own experience of sorting the envelopes over many years. Such a final system, with a machine carrying out the task, would then be referred to as an 'expert system', the robot merely copying the job previously done by a human expert.

As is the case in general, the robot sorting machine has just one specific role, and that is sorting envelopes. Even if we do allow it to learn, if all it can do is sort envelopes, then all it will actually do will be to learn to sort envelopes – nothing more, nothing less. In this situation we would not in fact really need or want the machine to 'learn' how to sort the envelopes, as we can directly program it to do the job exactly how we want it to be done. We humans have essentially learnt how to do the job over a number of years and we can program the robot with exactly that knowledge. In this way we can be sure that, barring unforeseen problems, the robot would do the job exactly how we wanted it done.

Even with a fairly rigorous, well-defined role for a machine such as the envelope sorter, it can be difficult for a human to know exactly why and how the robot is making its decisions, unless we are the human expert originally involved, the person who originally programmed the robot, or at least someone who has a good idea about the job. In the example above, only three rules were shown for the machine to check, but what if this list was 3,000 or 3,000,000? Would it then be so easy to know what the machine was doing? In the case of the robot envelope sorter it may not be necessary to have this many rules, but in other situations such numbers may be required. Also, if the rules stack up on each other, this in turn makes it much more difficult to understand, from the outside, exactly what is going on. Consider the following extension to Rule 1:

> If it is a large envelope *and* the address is handwritten *and* the county on the address is Berkshire *or* Oxfordshire *and* it is a Saturday, then put the envelope into box D.

As the problem itself becomes more complex, then it is often the case that a human carrying out the task will not get things right all the time, or may do things in a way that is not best, even though the human may have been doing the job for many years. In these situations it may well be extremely advantageous to let a machine learn how to do the job in a way which is better than any human has done it before.

Any task in which several things have to be done at the same time is difficult for a human to do, especially if the things are connected with each other. We can perhaps cope with two or three related jobs at the same time, but beyond that we are really struggling and often can cope only by doing one thing at a time and putting aside other tasks until later. We tend to be restricted to thinking in terms of two or three dimensions. Machines are certainly not so restricted; in fact, they are well suited to discovering multidimensional relationships and hence dealing with several things at the same time. In this way many machines are now used either to help humans in a particularly complex task, for example flying an aeroplane, or to replace the human operator altogether, for example in a telephone exchange.

Just as a machine can, by means of simple 'expert' decision-making, be set up to do specific tasks, so in the same way humans carry out many jobs in everyday life. If it is 7 a.m. then I must get up. If it is 10 a.m. then I must have a cup of coffee. If it is 5 p.m. then I must go home! It may be that when we say 'must' this might be dependent on a number of other things. For instance, rather than a cup of coffee I might have a cup of tea, and it might be 10.10 a.m., rather than 10 a.m., because I had to finish off a piece of work first. Essentially, though, much of our everyday and working life is run on fairly straightforward decisions.

Often we have learnt the decisions to be made, based on previously learnt decisions. If I want to live I must have some money. If I want some money I must have a job. If I want this particular job then I must get up at 7 a.m. Hence: if it is 7 a.m. then I must get up. But where can we trace this back to, this will to live? It is back to our genes, our initial program, our instincts.

Our own individual instincts and start-up of our basic character are dependent on our genes. We cannot choose our own genes; they are due to who our biological parents were, who their parents were and so on. Both our physical and mental conditions are set up by our genes; hence how we look, how we age, if we are susceptible to diseases, etc., will all have an initial start-up due to our genes. We can modify such things to an extent by exercise or taking certain chemicals or drugs. However, we each have basic tendencies which are difficult to change. Essentially, some people have a natural ability to run faster than others, or to lift heavier weights or to resist certain diseases.

A former research student of mine, Karam, told me, when he

was 27 years old, that he had never needed to go to the dentist even to check his teeth. He had no fillings, no tooth decay and no problems whatsoever. At the same time he did not need to bother much with cleaning his teeth. He had not learnt to look after his teeth, he was simply born with teeth of a much better standard than my own. Despite brushing my own teeth regularly several times a day, it has been a lost cause for quite some time and I regularly visit the dentist for repair work, which is not something I enjoy.

As well as affecting our physical state, our genes direct our initial mental state, our basic instincts, thereby causing us to know how to cry, breathe and sleep, how to acquire language and how to start learning. They also dictate some of our basic characteristics in terms of happiness and relaxedness, although it soon becomes difficult to say with any particular individual how much is due to his genes and how much is due to what he has learnt, even in the first few seconds of life.

Our genes, our initial start-up program, can have much more of an effect on our lives than we might wish to think. As one individual it is impossible to say, because we do not know *exactly* what characteristics we had at birth and *exactly* how we have been affected by our experiences. Occasionally, however, we hear of 'strange but true' stories in which twins have been separated at birth and have led completely separate lives in different places until they meet up again years later.

One such case[10] is that of the twins Jim Springer and Jim Lewis. Both were adopted by separate Ohio families when only a few weeks old, and grew up completely independently in different towns until they met again at the age of 39. On meeting, they found that they drank the same brand of beer and smoked the same brand of cigarettes. Both men had a basement workshop and had built a circular bench painted white around a tree trunk. When younger, both hated spelling but enjoyed mathematics, and both had owned dogs, which they called Toy. On leaving school both men joined the police, got promoted to the position of deputy sheriff, and left after seven years. Both men married and divorced women named Linda, then both married women named Betty, with whom they had sons, although Jim Lewis' son was called James Alan whilst Jim Springer's was named James Allan. Both men took annual holidays at the same time at the same Florida beach, although somehow they never met up. In more recent times they both took a multiple-choice intelligence test and answered the questions with almost identical answers.

We can simply put such phenomena down to odd quirks, and maybe that is true to an extent, but surely some of it, if not all, has been biased heavily by the gene make-up of the individuals concerned. The interesting point is that we do not know how much of each and every one of our own lives has already been predestined by our genes at our birth – if not to the exact detail, then at least in terms of a bias that exists throughout our life. Present-day research is uncovering more and more features that are dependent on our genes. Where will it stop?

Many things occurring in our lives we put down to random behaviour or pure chance. Often such things are caused by our interaction with other humans, and as a result we may learn to do things differently in the future, thereby affecting not only our own future but also the future of the people with whom we come into contact. But with other humans it is probable that we cannot predict what they are likely to do or say, simply because we do not know them very well. If we do know someone very well, to the extent of knowing how he will behave in almost every circumstance, we are very unlikely to be surprised by his actions, we will not consider his behaviour to be at all random and we will not feel that we will learn from his behaviour, although his actions may reinforce what we have already learnt. In the same way, if we knew *exactly* a person's gene set-up, in its entirety, we would know exactly his basic instincts and how he could learn new behaviours. If we also knew *exactly* his entire life experience, then we would know that person exactly, and would know exactly what he would do at all times. We would not then be surprised by any random behaviour by him; indeed, we would be able to predict what he would do in response to a particular situation, before it occurred. If we knew the same about everyone with whom we were to come into contact during a day, then the whole day would become, in terms of interaction with other people, completely predictable.

Of course we do not know the gene make-up of other people, nor do we know the exact life experiences of another person. The closest we get to this is, as described earlier, when we know a particular person well through being in close contact with them over a long period of time.

But is this all there is to a human? Are our brains simply deterministic and predictable? If I knew my exact gene make-up and my entire life experiences, could I not work out what I was scheduled to do and decide to do otherwise? Or would that decision itself be part of the schedule? Realistically it is extremely unlikely that we will ever be able to answer these questions. However, in Chapter 5, the operation of a human brain is considered in more detail, so we will leave the question for now.

All experiences in a human's life are important, and are unique in that they are witnessed through each human's individual senses. How we each witness a particular event depends, again, on our gene start-up program and our accumulated experiences. A football match between two teams can be seen as a good, happy match or as a bad, depressing match, dependent on which team one supports. So what an individual learns from an experience is really individual to them, and depends not only on physically how they witness the event but also their original gene make-up and their experiences to that point. I remember being impressed and very excited the first time I witnessed a vertical-thrust aircraft taking off. Others there were perhaps not so impressed, either because they had experienced the event many times before or because, due to their gene make-up, they were not really bothered about such things. It may well be, of course, that some others were even more impressed than me!

With a child it is often the case that we try to teach it certain habits by reinforcing and threatening. For example, we tell it to say 'thank you' when someone gives it something. Despite this, a child can take a very long time actually to carry out our required responses and may even never respond in the way we would wish. On the other hand the same child may learn to say a particularly bad swear word even though it hears it only once. Obviously the way in which the child is taught is critical. If we giggled and pulled faces if the child said 'thank you' it would probably learn very quickly that this is a good thing to do as it gets a good response!

It is sometimes said that children should not be allowed to watch violent films because they may learn from them and will themselves become violent. To an extent this is true, particularly if we adults have made a big commotion about a film, so children will watch it just to find out what the fuss is all about. This arousal of curiosity is very much a basic instinct, and is essentially what advertising is all about. However, in our genetic make-up we each have a tendency, or otherwise, to violence and anger, and through experience and learning we decide on how much violence is a part of our own lives and what we will do about it. A violent film may well teach a violent child new things to do, but unless the rewards are obviously positive, the child is unlikely to learn to carry out the particular violent acts.

Much more likely to have a serious effect are news items depicting violence, particularly in the same country as a viewer and particularly if the viewer associates himself fairly closely with the people carrying out the violence. In this case the message is clear. The person I am looking at is similar to me. He is carrying out a violent act. He is being shown on television and millions of people, like me, are looking at him. If I carry out the same violence, maybe millions of people will look at me as well.

It is through this thought pattern that inhibitions against behaving aggressively are reduced and eventually even removed. With the child coming to believe that violence is not only a typical and permissible way of obtaining results, but that it also rewards the offender. Couple this fact with that of the gradual desensitisation to violence, through frequent displays of violent acts in news broadcasts, and it becomes no surprise that today's children and adults are possibly far more violently minded than those in the past.

In this way we try, as adults, to decide what other adults or children can learn, by defining, to an extent, the environment in which they can learn. In the case mentioned here, we decide which television programmes are to be watched and which not. A supportive argument is that if children do not witness and do not know about a particular thing, such as a specific violent act, drug-taking, etc., then they cannot learn to do it themselves. Indeed, it must necessarily be true that if some individual does not know of some specific act, then he will not do it unless through gene make-up and experience of life he separately concludes that the act is possible.

Unfortunately, televising programmes explaining why certain things should not be done – that is to say, trying to teach people not to do the things – can have a very negative effect. Firstly, it makes people watching the programme actually aware of the things, which they might not otherwise have been. Secondly, if they become aware that people very similar to themselves have actually done these things, then no matter what those people now say about how bad it is, a human's basic instinct of experimentation may come into play and the people watching may well subsequently do the things in question.

The question of drug-taking is a good example of this. Should programmes explaining all about drugs, where they are purchased and how they are taken, be shown with a view to indicating how bad drug-taking is, particularly when some present or ex-drug takers are shown stating how bad the whole thing is and why they will never do it again? On the one hand, not showing such programmes is turning a blind eye to what is actually going on and what should, arguably, be discussed. On the other hand, it does bring drug-taking to the attention of those who might otherwise not have been interested in it.

In terms of a message for this chapter, it is extremely difficult to define exactly an individual human's learning environment, although this can be done, to an extent. It is not, therefore, realistically possible to know well what another unknown person is likely to do in a given situation.

At present the converse is true with most machines, such as robots, and therefore different conclusions can be drawn. Machines tend to be specific to one particular job, and so even if they are able to learn how to do that job, they cannot either learn from experiences other than those of which we are aware, or do things other than the task for which they have been prepared.

The word 'robot', although it is said to have come into use in more modern times through the work of Karel Čapek, originally described slave-like human workers in the fields. Such workers would do only what they were told to do, and would have essentially one role in life, namely to do the task which their master required of them. Their existence was very much dependent on their master's wishes. So it appears that there certainly are considerable similarities between such human robot workers and robot machines. Both are scheduled to do only one or a small number of tasks, to do so repetitively, probably in bad conditions and under the control of a 'superior' being. Both are, realistically, able to learn, but the superior being is very much in control of what they learn and can strictly control opportunities for learning.

Differences are that the human robots are all fairly similar in appearance, even being similar to their masters, whereas the robot machines are usually job-specific and their appearance depends on their function. Differences, or lack of them, in the mental make-up of humans and robots will be discussed later. However, it is worth considering the physical differences a little more.

As pointed out earlier in this chapter, it has been of considerable interest, certainly in fiction writing over many, many years, for humans to imagine a machine robot which is human-like, a humanoid robot. In fact it is not only in fiction; this interest stretches into the scientific community, and we will look further into the possibilities in the next chapter. Scientists have also shown a considerable interest in making robot machines which resemble animals other than humans.

A good friend of mine, Toshio Fukuda of Nagoya University, has perfected a chimpanzee-like robot which swings from branch to branch in different modes. The main interest here is the physical nature of the chimpanzee's arm movements and the ability of the robot to grab hold of one branch whilst letting go of another, thereby moving from one place to another.

In another vein, Shigeo Hirose[11] of the Tokyo Institute of Technology has worked on snake-like robot machines, which have multi-linked joints and are intended to imitate the biological movements of a snake. In this way the creeping movement of snakes on land is mimicked, as is their method of gliding.

As described later in this book, our own work at Reading has produced a number of hexapod (six-legged) walking robots which bear a resemblance to a number of insects, not only in the way they move but also in the way they send signals around their bodies. In particular the hexapods copy the movements of cockroaches, albeit on a different scale. The intention is not to mimic the walking movements of a cockroach, but rather to learn from what we know of a cockroach's movements in order to construct a robot walking machine. In this way the robot machine is in some ways comparable to a cockroach, and has been designed by taking on board some ideas from that insect. Nevertheless, it is no more than a robot walking machine, with its inherent advantages and disadvantages.

It is interesting to note, as Akiyama points out,[12] that when robot machines are animal-like, or seen to be small, friendly devices, 'We give them names, we want to stroke them, we respond to them not as machines, but as close-to-human beings.' My own experience is similar to this, but rather than 'close-to-human beings' it appears that 'animal-like beings' would be a better description. Certainly the response not only of children, but of grown adults to our own 'seven dwarf' robots, described later in this book, is that they are 'cute', 'sweet', 'fun', and I have heard comments such as 'Ooh, what are they doing now?' and 'They need a rest!'

Certainly humans do not usually treat heavy industrial robots in an animal, friend-like way, although we may give them a name. However, where robot machines are much smaller, are seen to be in need of our help, guidance and support, and, rather like a domestic pet, are felt to be less intelligent than ourselves and do not pose a physical threat, then it is extremely interesting how, as humans, we very quickly give them an animal-like status. This is particularly true if the robots are actually given physical looks which are, in some ways, similar to those of animals. Where a robot machine has such an appearance, then we relax in its presence. So, if a small robot is given a face with what appear to be eyes, does not make a loud noise, has fur (even if artificial) and moves around at a realistic animal-like speed, not too fast and not too slow, then we tend to treat it rather as we would an animal.

As humans we do appear to like the robot to respond to us when we are near, either by running away or coming towards us. We tend to speak to such a robot, even though it may not be able to hear us and though it might not respond to our voice even if it did. If, however, such a robot machine is made smaller, does not respond to our presence and does not look like an animal, then we tend to treat it rather like a usual robot machine or a toy!

Possibly all of this can be attributed to our experiences in life and what we have learnt. If we learn that fairly small, furry creatures with eyes are cute and fun to be with, then the fact that such a thing may be a robot machine does not immediately change our opinion.

It was indicated earlier in this chapter that even in a programmed robot machine, certain adaptive behaviours are possible. While the adaptive behaviours are concerned with sorting envelopes into boxes they may, or may not, be useful, but certainly they do not present us with any problems. When the adaptive behaviours are linked with small, animal-like robots and allow such a robot to change its mode of response, then we can often regard the response as being nice and good fun. Even with an industrial robot manipulator, it may well be that the robot merely adapts to a certain set of circumstances and changes its job function in a similar way to the envelope sorter.

When robot machines are small and animal-like or when they are carrying out a repetitive task for us, this is seen in a positive light. We need have no worries, even if the robot changes its action in response to some event or situation. But just as there are robot machines which we regard in a positive way, so there are also those that are felt not to be so friendly, especially when they are designed with the intention of directly harming humans or at least playing a military role.

The Surveillance and Reconnaissance Ground Equipment (SARGE) vehicle is based on a four-wheeled all-terrain frame and was developed jointly for the US Army and the US Marine Corp. It is a small and highly manoeuvrable scout telerobot which houses thermal imaging, day and night imaging and zoom surveillance cameras. Its sister vehicle DIXIE meanwhile uses dead-reckoning navigation – that is, it calculates in advance where it is going to go and then heads straight there – and triangulation, along with a video camera for positioning. These vehicles both roam around with no one on board, although military personnel can remotely view the terrain through the robot's eyes.

In a similar vein, the Pioneer unmanned aerial vehicle has the task of locating SS-21 Scarab missile batteries. It houses sensors specifically designed for the single task in hand. The same is true for the Loval Systems Patriot Advanced Capability (PAC-3) missile, which has an on-board millimetre wave radar in order to home in on a Scud missile target. Once fired, a PAC-3 will select and follow an enemy missile, exploding on impact. The PAC-3, therefore, makes a calculated decision on the target to be followed and will carry out its own interception policy based on the radar images it receives.

Now let us consider the completely automated 'Prowler' (Programmable Robot Observer with Logical Enemy Response). Prowler is a product of the American company Robot Defense Systems and was in fact originated in the early 1950s. It is about the size of a tank and in its original version was able to carry out a range of duties when under computer control. A realistic aim for the end of the century was that the Prowler would be able to observe the enemy and to carry out offensive action through its own adaptive behaviour. It was indicated that the Prowler could be armed with a 105mm cannon, Stinger surface-to-air missiles and a battery of M60 machine-guns. In terms of sensors the Prowler could, if required, detect body or engine heat and movement in a battlefield scenario, decide on whether the observed object is friend or foe, and act accordingly. In simple terms: 'If it is the enemy, then attack.' It would make further decisions on how to attack. In this way the Prowler could even take on low-flying enemy fighter planes and helicopters.

Is this surprising? Certainly not! It is merely making use of well-known technology. The only real complexity is in the adaptive decision-making behaviour, and yet all that is necessary is that this has the same basic requirements as the envelope-sorter;

indeed, the decision-making probably needs to be more complex for the envelope-sorter. But the Prowler, being a tank, is limited to battlefield operation. Technology has, though, moved on far beyond this in the last few years and we will look again, later in the book, at what is now possible with military systems.

So, what we are faced with, what is technically available now, is an autonomous military hunter which has been 'programmed' to select and destroy enemy targets. If we are on the same side, if we have programmed the hunter and switched it on, then that is in order. However, if we are the opposition, if we are being hunted, then what do we face? We face a hunter, maybe even several hunters, which will come looking for us. Once started on their quest we cannot rely on them to stop, they do not need to rest like humans, they probably will not make the same sort of mistakes as humans, and for sure they will respond a lot faster to us than we will to them. So we must try and get them before they get us!

But surely this is science fiction? Sorry, no it is not! It is not even present-day technology; it is technology that is several years old, particularly the adaptive behaviour, the decision-making.

This chapter started by looking at fictional concepts of robot-like creatures, all of which seemed to be physically similar to humans, but either not so intelligent as humans or under human control. Yet we have ended with machines which are not like humans, which may not be as intelligent, but which nevertheless have a distinct advantage over humans. Not only will they not go away, but rather their advantage will increase as time passes!

Chapter Four
Humanoid Robots

In the last chapter we examined some of the ideas humans have about robots, and the long-standing desire, in myth and fiction, to build a humanoid robot which would look, act and behave like a rather stupid mechanical human. However, humans have built up another concept of what a robot is, this time based firmly on reality, and that is the standard robot arm manipulator which is used on many production lines throughout the world. In this case the robot is usually employed for a specific task, for which it is programmed, and it repetitively follows a scheduled sequence of events. The robot is switched on and off by, and is under the control of, a human, in that it has been programmed by a human, who therefore knows exactly what it will do, unless a fault occurs.

The standard production-line robot is a long way from being like the humanoid robot ideal. It is essentially just a fixed mechanical arm, a physical extension of the capabilities of a human. It may or may not be stronger than a human. However, it is invariably able to carry out its required task accurately and repeatedly, even operating 24 hours a day, seven days a week. So these are some of the advantages of such an arm in comparison with a human carrying out the same job. In fact, even with only a limited number of hours to do a job repetitively, humans often make mistakes and perform poorly, especially on a Friday afternoon.

Robot arm manipulators are even now being employed anew on tasks previously carried out by humans. Many production lines are becoming completely automated, with human intervention only at a decision-making level or to deal with machine failures. In most cases the initial cost of purchasing and setting up a manipulator for a task is fairly high. However, it is often possible to recoup the costs over a two- or three-year period from the savings made through not employing a human to do the same job. The human often also demands good working conditions, with adequate heating, lighting and canteen facilities. None of these are necessary for a manipulator. In fact, some repetitive tasks are fairly repulsive, or actually dangerous to humans; for example, dealing with animal hides in a tannery or working with nuclear reactors. Not only is the requirement for pleasant working conditions reduced when a manipulator is used, but humans usually demand much higher salaries to compensate for working in unpleasant conditions, making it even more sensible to use a manipulator if possible.

There are, of course, other types of robot in practical use today, although they are by no means as widespread as industrial robot manipulators. One example is that of toy robots or imitation figures such as 'Dingbot', which is described as 'crazy and fun-loving'. Each Dingbot is several inches tall and hurtles about on small wheels. When it encounters a wall it turns around and makes strange 'roboty' bleeping noises. For eyes, it has lights which flash periodically, and it has small arms which can be moved into position by hand. Meanwhile, Omni 2000 is used to carry light objects, such as drinks and food. It, therefore, spans the gap between toy robots and domestic robots; that is, robots used around the house.

It might have been expected that domestic robots, moving around on legs or wheels, with hands and eyes, and designed to perform domestic chores, would be more widespread than they are. In fact, many domestic tasks, particularly in the kitchen, are carried out by machines, though not robots. We have machines for washing clothes, machines for washing dishes and machines for cooking. Machines such as these are perfectly well accepted in the home, but are often not immediately thought of in terms of domestic robots.

Other robot forms are less widely encountered; for example, bomb disposal robots, which are remotely controlled for work in obviously dangerous environments, or vehicular robots. It is interesting that standard automobiles or locomotives are not considered to be robots, yet they are machines under direct human control. If, however, an automobile is required to move and direct itself to follow a white line it is then, somehow, regarded as a robotic vehicle. The bomb disposal robot is controlled in much the same way that we control an automobile, or a go-kart at least, yet because we are remote from it, it is termed a robot. Indeed, we have strange definitions as to what is, and is not, a robot. This book is concerned with machines in general, whether or not we call them robots.

Robot-type machines come in a number of forms, electrical being the most common, with electronic or computer-based control. More power can, however, be obtained by use of pneumatic (air) or hydraulic (water or liquid, e.g. oil) techniques. It can be argued whether cranes, powered ladders and extending platforms, which are certainly machines, are or are not robotic devices. One thing that is clear, however, is that extending cranes, bomb disposal robots, washing machines and locomotives all have a common link in that they neither look nor act like humans. They are helping us to do a particular job, in one way or another, but they are not about to replace us in our entirety. What then of robot machines that do look like and act as human replacement parts?

In fact, we already use machine parts to replace components in human bodies. Many human organs, bones and joints are replaceable. Hip joint replacement is a common operation, with heart replacement being a more recent introduction. Mechanical lungs have also been around for many years, as have 'artificial' mechanical legs, and kidney machines. Significantly, the brain is, at present, holding out as not being replaceable, and often it is required in order to make the particular mechanical replacement work correctly.

An important point is that a machine which replaces a human in carrying out a human task usually does it more effectively than a human, but will most likely not be at all human-like. Simply because a human has been doing a job for many years does not mean that it is the best design to actually perform the task in question. In fact the opposite is probably true, in that the advantage of humans, at present, is our versatility. We can do many things, whereas machines tend to be job-specific. Just about everything we do, therefore, can probably be done better by a machine which is specifically designed for the task. It may be, though, that the task itself is modified in order to suit the machine, as with a washing machine: we used to wash clothes differently before the first washing machines were built.

Many roles carried out by humans will appear to be difficult for machines, particularly when a direct comparison is made with how humans currently do them. As an example,[13] consider posting envelopes through letter boxes in house doors, as is the custom in the UK. Potentially this job is very difficult for a machine, not so much in terms of actually putting an envelope into a box, but rather in locating the correct door, at the correct house, and moving up the path to the door without hitting objects, such as animals, parked cars or brick walls.

However, a task can quickly be modified to suit the machine. Simply putting a mail box for each home at the side of the street, which is the custom in the USA, makes the overall task much more amenable. Furthermore, a unique identification tag can easily be placed on the box, so that the final solution of a robot delivery service becomes a realistic proposition. It is simply a case of how much we, as humans, are prepared to adapt quickly to our tasks being taken over by machines.

Thinking about delivering mail, whether by human or by machine, is in itself rather looking back into the past. Handwritten or typed messages on paper have been around for many years and are, of course, now being seriously challenged by electronic mail. This computer-based method of sending messages could ultimately avoid the need for any mechanical mail service whatsoever. All the emphasis is placed on messages being sent from one computer terminal to another via a network. There are no physical mail boxes, no physical delivery, no problems with hitting physical objects. So, in some respects, it is not even sensible to consider whether or not a machine can replace a human in doing a task. There are times when the whole technological base moves on and leaves the old methods behind.

There is a school of thought which says that we humans have something extra. We are biological systems with electronic, chemical, pneumatic and hydraulic functions. Nothing, no robot, no machine, is ever going to be like a human. We are not only more intelligent than anything else, with distinctly different and superior mental properties, but physically we can do things, such as walking, running and moving our hands, in a way that no machine can do. The mental comparisons will be looked at further in the chapters which follow. For a moment, let us look here at how close machines are, physically, to human movement and actions.

A group at Hull University has developed artificial muscles, in which polymer contraction and expansion is affected by applying chemicals. Forces similar to those produced by a human hand have in this way been originated by the artificial muscles, to achieve similar gripping strengths. A Japanese group is working on roughly the same principles, although electrical rather than chemical stimulation is employed. The muscles can, in each case, be operated under computer control, by sending a stimulation signal to the muscles so that they operate correctly.

The human hand is an extremely complex entity and it is very, very difficult even to get near to copying it in an artificial way. However, it is relatively easy to produce an 'artificial' two-fingered gripping device, along the lines of a crab's claw. Of course, when a requirement for some sort of sensing is introduced – in order to assess, for example, how much pressure is being applied to an object by the fingers – together with a requirement for more fingers, possibly five, operating collectively, then the problem becomes much more difficult.

At present, machine hands fixed to the end of manipulator arms are usually two-fingered grippers, or else special-purpose grasping mechanisms for the task in hand. It is worth remembering that many jobs do not require, or are not best done by, a human type of hand; for example, suction pads for lifting glass or strong rods for heavy objects. Even humans use various 'gadgets' when operating on things for which our hands are not well suited: spoons, knives, fishing rods, tennis rackets, etc.

Nevertheless, a considerable amount of research has gone into robot hands which are, in some way, aimed at mimicking how a human hand works. At Chuo University in Japan, the XI hand has a total of 384 sensors on five fingers so that when the fingers come into contact with an object they can obtain an indication of its shape, size and even texture. Like many robot hands, the XI hand operates in 'stand-alone' mode, in that it is not designed to be connected directly on to the end of an arm; it is merely there to prove a point, as a research tool.

Lifting glass with suction pads is a good example of where a sufficient amount of grip must be applied to do the job. When humans use hands, we need to apply the correct amount of force to an object, possibly even a twisting or spinning force, to apply a sufficient grip, to feel any slipping, to sense temperature differences and to establish an object's texture, shape and size. Human hands deal with an enormous range of things: paying extortionate amounts of money for this book, knocking on a door, steering an automobile, knitting, handling liquids or sticky, slimy substances. Our hands are extremely versatile, so that approaching this versatility with artificial machine hands is very difficult.

Some gripping fingers have been constructed which are not too complex to operate and which can be controlled to previously programmed limits. The actual fingertips can be made of a spongy substance, so that breakages of the objects held are not frequent, or they can be made of a sticky substance if slippage is not desirable. Fingers can also be shaped around an object in a number of ways. One possibility is demonstrated by Omnigripper (Imperial College, London), which contains rows of pins on its fingers. When holding an object, the independently operating pins mould completely around it, thereby obtaining a direct picture of its shape. The same type of results can be obtained with fingers full of granules. When taking hold of an object the granules are loose. However, once the object has been gripped the granules are held in place, possibly electromagnetically, and the fingers take on the object's shape.

In order to be anything like a human hand, a robot hand needs to be multi-fingered, dextrous and flexible. The use of a hand, for a human, also involves selecting the correct posture for the appropriate problem, predicting what is expected to confront the hand before a task is carried out, and dealing with a variety of unforeseen circumstances when the task is being undertaken. Therefore, not only are the physical characteristics of the hand important, but so too is the way in which the hand is controlled.

A number of relatively versatile multi-fingered robot hands have been produced; for example, the Utah/MIT hand,[14] the Belgrade/USC hand,[15] the JPL/Stanford hand[16] and Reading University's own CybHand.[17] Each of these involves an actual, physical robot hand, rather than a computer simulation, and each has certain properties. The Utah/MIT hand is perhaps the most widely reported, this being a four-fingered (that is three fingers and a thumb) hand, each finger consisting of four joints. The joints are individually operated through wire connections, rather like tendons. Overall control of the fingers is by a central computing unit.

The Belgrade/USC hand consists of four fingers, each with three joints, and a rather stumpy thumb. Each pair of fingers is operated by one motor. However, many of the hand control functions are local to the hand; it is an intelligent hand. This hand is equipped with sensing devices to detect contact, pressure and slip, with the aim of locating control decisions at the hand. A new version is, I understand, under construction and it is promised that each finger will have its own driving motor.

CybHand is, like the Utah/MIT hand, a four-fingered device, each finger having three joints which are moved by motors directly driving them. Each joint has its own motor. The hand has been successfully used to grip a range of objects, and has been employed for the fingers to *learn* how much pressure to apply to objects. The hand has performed perfectly on many occasions, holding a wide range of objects. However, on one occasion the local newspaper wanted to run a story on the hand and sent a photographer round to take some photographs, so we took much time to ensure that it was working faultlessly. When the photographer was with us we gave the hand a pen to move around – no problem; a sponge – no problem; a china cup – the hand promptly dropped it and it smashed into pieces on the floor. Nice photograph!

Though mechanical, in many ways these hands resemble human hands and indeed in their movements appear to be extremely hand-like. Also, in terms of their operation, wires moving each of the joints are rather akin to tendons and are generally controlled individually. A lot of information can, therefore, be gained on the problems faced by human hands. In truth, such robot hands are far from being anything like human hands in terms of their actual operation. Mathematics are thrown at the hands, with extremely complex and detailed physical and mathematical models being formed of each hand's operation. Is it correct to drive each finger joint separately? Try moving any one of your own finger joints on its own, individually. I bet you cannot do it very easily, if at all. Complex mathematical relationships produce hands which are difficult to control and which can carry out only a limited number of tasks. It is said that the Utah/MIT hand does not hold a screwdriver with ease, and of course CybHand has problems with cups!

Material for use as an external cover on the robot hands is still a long way from being as versatile or as robust as human skin. Touch sensors are generally placed at a few specific points, often simply on the robot hand's fingertips. It has also proved extremely difficult to equip a robot hand to cope with the range of tasks that a human hand deals with.

All of this means that in terms of copying the make-up and operation of a human hand, robot hands are generally not as versatile, and are also extremely difficult to control. What does this mean? Well, probably it will be many years, if ever, before artificial hands are comparable with human hands. Unfortunately, this means that in the near future those humans who have had a hand amputated and are presently fitted with the dual hook system, the most common replacement, are unlikely to be fitted with an artificial hand which is roughly comparable with the human one.

It is worth taking stock for a moment and looking at how humans control their hands. Do we use a lot of mathematics, numerous complex physical equations, in order to work out planned movements and gripping schedules? Well, I certainly don't! Do we predict what type of pressure to apply to an object before we touch it or pick it up, based on extensive mathematical descriptions of the structure of the object? We probably do make some prediction of how much to squeeze the object, but this will most likely be based on our experience of similar objects that we have previously dealt with, and not a mathematical description. Throwing mathematics at the hand in order to control it is, therefore, probably not the best approach if we are trying to achieve an artificial hand which is controlled in the same way as a human one.

Even if we did manage to control the robot hand in a human-like way, each hand is still physically unable to deal with the wide range of things a human hand can deal with. But would a robot ever need to carry out such a range of tasks? After all, it is a robot, it is a machine, it is not a human!

Now close your eyes. Apart from the obvious problem of finding it difficult to read this book with your eyes closed, try moving your arm around at your side. How do you know exactly where your hand, or arm, is at any particular time? Presumably you can still 'feel' roughly where it is through your nervous system giving an indication of anything touching it, how much your muscles and tendons are stretched and how much your joints (elbow, shoulder and wrist) have moved. With your eyes open, however, it is very easy to follow your arm around visually and to couple together the information from your eyes and nervous system in order to get an exact picture.

With your arm held out to your side, how far exactly is your hand away from the ground? What exactly is the angle of your elbow joint? Now move your hand. How fast is it travelling, exactly? I, for one, would only be able to make a very rough guess at any of these answers, unless I was connected up to a complicated and sophisticated measuring device. Yet with robot arms, these values are exactly the things that are measured. The robot arm control system does not in general get a 'feel' of the arm position from an inbuilt nervous system, neither does it usually get a visual picture for confirmation, although this is possible. Rather, a mathematical calculation must be made, based on the various measured values. In the same way as for a robot hand, a robot arm is usually controlled by throwing a lot of mathematics at the problem, which, again, is certainly not something that humans do in controlling their own arms.

How do we know the way in which to move our own arms? Isn't it through a combination of our initial instinct, our gene program, and what we learn by experience, usually when we are quite young? But humans cannot only move one arm around, we can move two, often in a coordinated, cooperative way. For many tasks in life where hand coordination is useful, we learn fairly quickly how to move both arms and both hands at the same time. One example is driving an automobile, another would be lifting heavy objects. Often we hold an object firm with one hand whilst operating on it, in some way, with the other. Some other tasks requiring hand coordination are much more difficult. Juggling balls is a good example, and can take some considerable time to learn; indeed, many people may never learn. In each case, however, no physical change occurs to enable us to carry out the task, merely a mental one in that we learn how to do it.

Just as the control of a single robot arm, acting on its own, is difficult, so operating two arms cooperatively is even more problematic, although if a particular two-arm arrangement is designed for a specific task, then this is certainly possible. The key is that as long as no flexibility or learning is allowed, all eventualities can be previously accounted for and the robot arms programmed accordingly. In fact, in the early 1970s the Sterling Detroit Co. produced a coordinated two-arm arrangement to deal with steel castings. In this case two arms were necessary to deal with all of the manipulation requirements in a working environment which humans found undesirable, principally due to the heat, for which they therefore required high salaries.

KUKA, a German robot manufacturer of industrial production robots, has put together a range of coordinated robot cells. In one arrangement, four robot arms work simultaneously as welding robots on such tasks as making combine harvester mainframes. Clearly, using four robot arms keeps production time relatively low and is cost-effective. During the welding sequence the arms move around in proximity to each other. However, it is necessary to ensure there is no arm contact. This is readily achieved because the entire welding sequence is previously programmed.

71

It is interesting that quite a number of robot arms – four in the KUKA case – can work in a coordinated way, whereas for humans it can be extremely difficult to get more than two arms working in a coordinated fashion, other than on simple or relatively slow tasks. This is, of course, due to the need for more than one human, and therefore more than one brain-driven control system. The problem is then more of brain coordination than arm coordination.

For coordinated robot arms, where flexibility of movement or even learning of movement is allowed, collisions between the arms usually need to be avoided. Robot arms are not, in this sense, as robust as human arms. However, in order to ensure that robot arms do not collide, either a no-go volume can be arranged for each arm – that is, a region where the arm is not allowed to go under any circumstances – so that an arm never gets into a position where it can collide with another arm, or else the arm controller must be aware not only of where its own arm is but also of where the other arms are. In a two-arm system this is simply a case of the right arm knowing what the left arm is up to.

Just as human arms, and even more so hands, have evolved into extremely complex devices which are difficult to replicate artificially, so our basic means of moving around, walking, is also relatively sophisticated. In order to consider how far robotic machines have been developed so that they can walk, it is perhaps best to look first at the easier task of walking on a number of legs, rather like mammals with four legs or various creatures with six, eight or more.

In fact six legs seems to be a pretty good selection, in that this choice does not involve too much in the way of coordinated control. The coordinated control of a 20-legged robot is a relatively difficult task, unless several of the legs are coupled together, thus reducing the control problem. Six legs is not such a difficult problem from a control viewpoint. However, it does offer the distinct advantage of rapid yet stable movement.

With only two legs, the human model, simply standing up needs an active balancing operation, which is something that humans learn to do. Walking means that this active balancing must be continually in operation, in that humans effectively walk by a series of controlled falling-overs!

A six-legged, insect-like robot with all legs on the ground should be able to stand still nice and steadily. There are no tremendous balancing problems unless the design is particularly bad. Movement can then be achieved in one of a number of reliable ways. One obvious possibility is to move one leg at a time, keeping the other five standing. A more representative option, however, is to use stable tripods. In this way, at one point in time the left front, left rear and right middle legs remain standing, thus forming a stable tripod. The remaining three legs can then all move. Once this move is complete the right front, right rear and left middle legs remain standing whilst the other three legs move. The (robot) insect can thus move around without having to worry too much about balancing.

Quite a large body of research now exists[18] relating to insect-like robot movements on six legs, not only through the actual basic design but also in terms of controlling the device to sense obstacles and move around or over them. In particular it is fairly straightforward to sense when a foot is touching the ground, and when not, which means it is not only easy for the insect to detect and avoid holes but also for it to adjust its movement over an uneven surface.

At Reading we have a number of six-legged walking robots, in particular 'Walter', and our most recent addition, 'Elma'. Elma not only has a tidier body than Walter, she is also much more intelligent! The key to these robots is, however, not so much their mechanical design but rather their means of control and decision-making. We will meet up with them again later in the book.

Four-legged (quadruped) movement, such as that exhibited by horses is, I believe, more complex in terms of gaining a firm understanding of how to achieve not only walking, but also trotting, cantering and finally galloping, in which all four legs are off the ground at the same time. Again, for a robot quadruped it is simplest to retain three legs on the ground at all times, whilst moving the one remaining leg, although this is a relatively slow form of movement. The other possibilities, of either moving both front and back left legs and subsequently front and back right legs, or moving front right with back left and subsequently front left with back right, involve delicate balancing techniques, thus adding to the complexity of the control problem.

However, at the University of Abertay, Dundee, a four-legged robot, called Boris, has been developed, which moves by pushing off with its back two legs and landing on its front two. This means that at some points the whole robot has its feet off the ground, and it must therefore have the ability to land in a balanced fashion in order to take its next step. While at Reading University, one of the 1998 student projects has developed a robot cat called Hissing Sid. Although the cat moves by lifting only one leg at a time, it incorporates an active tail, which is used as a counterbalance to the leg off the ground.

Four-legged robots are fewer on the ground since they do not have the control and stability advantages of six legs. Nor do they have the humanoid robot appeal that two-legged robots have. Nevertheless, common problems exist insofar as movement generally requires a number of detailed mathematical calculations, based on fairly comprehensive models of the four-legged robot and its characteristics when in motion. Again, the question arises, is this the best way of going about it? The animal world is full of examples of four-legged motion. The cheetah is the fastest of all mammals, reaching speeds of up to 70 m.p.h., but it is difficult to imagine that it mentally runs through a sequence of complex mathematical equations every time it moves a leg!

Before we learn to walk on two legs, of course humans learn to crawl on four legs. In this case we actually crawl by moving, in the first instance, one of our limbs at a time, although later on a diagonal pattern emerges. In this way crawling is a fairly stable operation, not too difficult to control.

Marc Raibert's hopping machines[19] require no cooperative leg movement whatsoever in the single hopper case. However, the leg, which consists simply of a metal cylinder which bounces up and down on a spring, is inherently unstable. So, if it is not hopping, it falls over. The machine balances by rotating as it springs and is fun to watch, although it appears to have a limited practical use. One problem with the hopping machines so far witnessed is that they have cables leading off the hopper to supply power and others leading off to provide computation. I am not aware that a completely stand-alone (or should I say hop-alone) single hopping machine has yet been built, with its power supply and computational abilities on board. I would be happy to be informed otherwise.

Human legs are multi-jointed, and much is to be gained simply by looking at the operation of one leg in terms of how the joints combine in order to achieve legged motion. A research student of mine, Nigel Archer, investigated the effects of gravity on the ankle and knee joints in a motor-driven robot leg he had constructed.[20] The leg, commonly known as 'Nigel Archer's leg', is now used for teaching, where it is shown how very simple methods can be used to control single leg movements.

The problem is made much more difficult when an attempt is made to construct robot machines that walk using two legs. The problem of standing on two legs is fairly easy to achieve. However, once the arrangement is set off balance or, worse still, one foot is removed from the ground, the situation is much more difficult, and rather complicated sets of mathematical equations are employed in order to achieve some semblance of walking. A number of Japanese groups have perhaps achieved most in this field so far, with the first 'Wabot' device, a set of 'Biper' robots and, most importantly, the Honda P2 robot. Wabot 1 is a hydraulic system with large feet shaped like plates, and several joints (ankles, knees), including a pelvic joint, which allows a leg, when moved, to head off in a different direction from the other standing leg. In this way Wabot 1 can turn corners and move around objects.

Biper, meanwhile, looks more human-like and is able to climb stairs, albeit rather slowly, by bending its knee joints and continually balancing itself. Biper uses electric motors for joints, thereby providing direct power; that is, the motors are actually positioned at the joints rather than being applied via tendon-like cables.

It is the Honda P2 humanoid robot however, which represents a distinct step forward in the development of two-legged robot walking. It was released during late 1996–early 1997, is very human-like in construction and is of approximate human-size, at 180cm tall and 210kg in weight. It is battery powered and although it moves autonomously, a radio control connection can be established to change its direction if it heads for trouble. It has the ability to make decisions according to terrain conditions, can turn, walk both up and down stairs and negotiate slopes. It even has the capacity to push a cart or tighten bolts. It maintains its balance autonomously, even if pushed, and if the pushing force is really large it keeps its balance by sitting down.

Robotic walking aids for the disabled are also of interest. In the Reading cybernetics department we recently developed a walking frame which would initially provide a lot of additional power to help a person learn to walk again, after a car accident, for example. As the person gains more strength in his or her legs, so the frame has less of an effect until, when the person is able to walk reasonably well, the frame takes no further role and can be removed.

A walking machine from the University of Wisconsin is more advanced. This arrangement is based on hydraulics and contains hip and knee joints. A person who is able to walk climbs into the machine, which is then 'taught' to walk, climb stairs, sit, stand, etc. by that person. Subsequently a person who is unable to walk can be substituted for the original person, and the machine will carry out the walking actions it has 'learned'.

With two-legged robot machines a number of problems arise. The first of these is how to get power to the legs. If the power source is included in the machine then it has a dramatic effect on the second of the problems, which is balance. The higher the power source is placed, the worse the problem becomes. The alternative is, of course, to remove the power supply from the machine altogether and simply run cables to provide the power. A second problem is that of manoeuvring and navigation. Certain information can be obtained about the outside world by detecting where there are obstacles and where it is not possible to travel. However, this needs to be interpreted and used for planning courses of action; for example, preparing for a step up or down. These planning features are in fact common to the robot hand machines described earlier in this chapter.

The importance of legged motion is paramount, in that, it is said, 90 per cent of the Earth's land surface consists of rough terrain, or at least terrain which is not readily accessible for wheeled vehicles. In many cases, in fact, the terrain is also not readily accessible for legged machines or humans. For machines to be able to access a large proportion of the Earth's surface, therefore, it appears that some form of legged motion would be necessary, although it may be that six rather than two legs would be more suitable.

In this chapter so far, the picture that has been given suggests that, if we were trying to construct a complete humanoid robot, we would encounter a number of problems. Although we might manage to construct a couple of good arms, it would be difficult to control them in a coordinated, cooperative yet flexible way. Although we might manage to put a couple of hands on the arms, actually getting the fingers on these hands to feel things and operate in a generally coordinated way would be difficult. We could then build a body and put a large power supply inside and the whole thing would sit on a pair of legs, which could be used for the robot to move around. Realistically, although several of these features have been achieved individually, an overall humanoid robot in a general physical sense has not yet been accomplished, although the P2 and its successors, the P3 robots are well on the way.

The humanoid robot thus far described is missing a number of things. Sex organs have been omitted. If you have purchased this book on the grounds that it will be hot and sexy, then I am afraid you have been sadly misled. We will, however, look at robot reproduction later in the book. For the moment, though, our humanoid robot will have to remain devoid of sexual organs; a celibate robot.

Another missing feature of our humanoid robot so far is a 'head', with all of its different facilities. The brain, the central processing and control unit, is dealt with in considerable detail later; indeed, it is a critical feature of the whole book. Other than that, a human head is littered with communication, survival (eating, drinking) and sensing devices. Instead of eating and drinking, a robot's means of survival is provided by a power source such as a battery, which simply needs charging. For machines, communication and sensing are usually very different than for humans. Communication will be covered later, but it is worthwhile looking briefly at sensing here.

We considered touch and pressure sensing, to an extent, earlier in the chapter. It is fairly straightforward for machines to be aware when contact is actually made with an object, merely by means of a switch that closes when touched. A number of methods then exist to find out how far away an object is. One method is for machines to send out a very high-frequency sound signal, an ultrasonic signal, which will bounce off an object and come back. If no object is present then the signal will not come back. This is the 'ping' that submarines send out, and it is also the method bats use to detect something nearby. We can measure the time the sound signal takes to bounce off an object and come back to us, and from that, calculate the distance of the object.

Similar principles can be used with laser signals, although it is more difficult to get an accurate measurement of object distance due to the higher signal frequency. This type of signal is better for longer distances. For even longer distances the same method can be used with electromagnetic signals, called radar, but this requires large equipment which is difficult for robots to carry. These techniques of finding out how far away objects are may be suitable for machines, but not for humans. Ultrasonic signals are even suitable for some mammals, but they are not used directly by humans.

The main human sense is our visual system, this being in stereo, with two eyes from which we achieve a concept of object distance. Our brains firstly process the image obtained through our eyes and then try and make sense of it and put it into context. We learn to recognise certain things – objects, people, food, places, etc. – often on a purely visual basis, and simply a visual picture can result in immediate learned responses from us. Machine vision systems are at present not really on a par with human systems, not because of the camera type employed but rather because of the processing and understanding of the images obtained.

The same conclusion is really true of understanding sound signals, in that although technically there is not really a problem in capturing a sound signal, getting machines to *understand* different sound signals is difficult.

A group at Warwick University has produced an electronic nose that detects a range of different smells, but this too falls some way short of the performance of a human nose, being able to distinguish between only a small range of smells. Artificial or machine tasting systems are even further away from human performance. As with vision, in each case the approach has usually been to throw mathematics at the problem in an attempt to find a solution; to analyse a smell in terms of its different parts, to analyse a visual image in terms of drawing lines on the image.

From all of this, the only possible conclusion is that at present we are a long way from a complete humanoid robot; that is, a robot machine which looks like a mechanical human, which moves like a mechanical human and which has a range of senses which are roughly comparable to those of a human. So is the story in the second chapter wrong? Surely it cannot be possible for machines to take over the world if they cannot even perform the simple things humans do?

Well, I am afraid that such a conclusion would be wrong. Machines are very different from humans. But we must not fall into the trap of assuming that because they are not the same as humans they are, and always will be, inferior. Many machines have been originated to help humans in what they do, while others are intended to replace humans in unpleasant tasks. A third category of machines, however, has been originated to do things that humans cannot do, or had not even thought of doing before the machines came along; these can then allow humans to do other new things.

So what sort of things can robot machines do?

As we have seen, robot machines have been employed in industrial environments, often to replace humans in boring, hazardous or unpleasant jobs. Usually the machine has several advantages in that it is often faster, more reliable, more accurate, much cheaper than a human, and not likely to suffer from Friday-afternoon blues. Such machines have been used on production lines now for 30 years or more, carrying out tasks such as assembly, welding and paint spraying. In the first instance they were, and in many cases still are, programmed via a lead-through sequence by an operator, each position and action being carefully checked during the teaching sequence, so that subsequently the machine can follow exactly, time and time again, the sequence it has been taught; a programmed robot.

But it was not long before robot machines were allowed some sensory powers, one example being simple visual information through a camera. This enables a 'pick-and-place' robot to decide which object to pick up and where to place it, dependent on which object appears in front of it. COMAU, part of the Fiat Group, has a number of robot systems with 'connected' visual systems. One arrangement is for the automatic assembly of automobile wheels. The machine, as a whole, must check the wheel hub on an automobile as it passes on the production line. A wheel, which is held by the robot, is then aligned so that the bolt holes line up. Finally the machine puts on the wheel nuts and checks for the correct tightness by means of further sensors. This example is typical of many.

Certainly for most industrial production companies, the key drive over the last 20 years or so has been to automate production in order to stay in business. If the company does not do this, competitors who have already automated will make their product much more efficiently, reliably, accurately and cheaply. In order to compete, automation has been a necessity.

Some robot machines have used visual systems to enable them to operate in potentially dangerous areas. GEC has a snake-like robot which winds its way around nuclear reactors, with a lighting and camera system attached. This enables one operator, sitting at a remote 'safe' site, to operate on the reactor, via the robot. The same type of device is used for bomb disposal machines and even on underground mining robots. Here a human controller can remain on the surface and is able, by moving his own head around, to cause the robot camera below ground to move around in sympathy. The mine can, therefore, be inspected from a safe, remote distance. This type of robot operation is called 'teleoperation', in which the intelligence or decision-making of the robot is still carried out by humans, but at a distance.

The rate of introduction of robot machines for automation and production has been tremendous, to the extent that modernisation of facilities is a perpetual requirement. This can even mean, in some cases, a complete change to a new set-up, in order to take on board the technology that has become available. Peugeot has recently upgraded its production plant in Sochaux, France, by replacing 200 robot welding machines with 91 new versions. Formerly the plant was able to produce a maximum of 1,230 Peugeot 405s per day, whereas now it is capable of realising 500 in just one hour. Many relatively difficult tasks have also been automated. Often these are tasks which 20 years ago were felt to be beyond automation. One example is the positioning and fixing of windscreens, both front and rear. The location problem is overcome using laser sensors.

One result of the employment of more and more automation and the use of robot machines is the gradual, but certain, removal of humans from carrying out any role in the entire production cycle. In many cases already, the full production shop floor is machine-based. Humans retain roles such as decision-making, in terms of which machines are working. They also carry out maintenance, to keep the robots running, and perform quality and error checking, to make sure the machines are doing the right thing. So here, none of the actual production is being carried out by humans. In fact, many of the remaining roles are also gradually being reduced by automation: for example, machines are being built which indicate what the problem is when a fault occurs, and dictate the best time for maintenance.

The drive is still on for more profits, and hence even more automation will be employed in the future. In many cases the aim is now for a product to be made round the clock, without human intervention, without failures and with a minimum of down-time. Fully automated factories are, therefore, rapidly becoming a competitive necessity rather than a dream for the future.

The manufacturing and production sector has obvious financial incentives for the employment of more machinery. However, there are so many other tasks, which have traditionally been considered to be human strongholds, that are now better done by machine. In some cases it is clearly just a matter of time before humans are replaced altogether. Shoe and clothing manufacture are good examples.

So what else can machines do? Well, we know that they can play chess, presently to the level of Grand Master, and as discussed further in Chapter 6, in May 1997 Deep Blue, a computer chess player, defeated, over a series of games, the human world chess champion, Gary Kasparov. It was in the 1980s when machines surpassed humans at backgammon. It was also in the early 1980s that a robot machine (Cubot), complete with fingers, was able to solve a jumbled Rubik's Cube, on its own, in less than four minutes. There are also robots that can play snooker and ping-pong, and at Reading we have one robot that can throw and catch a ball and another that can operate a yo-yo! Indeed, it is often the case that things that humans believe need a lot of talent can often be done much better by a fairly ordinary robot with little in the way of decision-making or sensors.

One excellent example of a not-so-ordinary robot is Waseda University's Wabot 2, which can play an organ or piano, and indeed has appeared with symphony orchestras. It has fingers to press the keys, feet to operate the pedals and can read sheet music by means of a camera vision system. It can, therefore, read and play any piece of music put before it, and one wonders how long it will be before new music is specifically written for robot machines, on the grounds that it cannot be played by human musicians. As someone who can play no music, other than an occasional whistle or hum, I used to be very much impressed by people who could simply sit at a piano and tap out a tune. Wabot 2 made me think again!

The United States Postal Service has designed a robot for cleaning washrooms. The robot cleans the toilet inside and out, in its entirety, lid, pan and seat. Indeed it gives a better performance than that obtained from human cleaners. Electrolux now manufacture a small vehicle which automatically vacuums around the home, operating over different carpet piles. It is referred to as a robot vacuum cleaner, although it looks more like a flying saucer on wheels. Meanwhile the German automobile company BMW has built a robot petrol pump which operates by reading information from a unit under the automobile. This tells the robot where the filler cap is on the automobile and the type of fuel which must be used. Laser scanners then guide the robot into position, at which time it opens the petrol flap, twists off the cap, puts in the nozzle and pours in the fuel. Automobiles can be serviced in this way in less than four minutes.

There is a robot[21] which changes shape and repairs itself using various-sized individual motorised cubes which slide over each other in any direction. It can transform itself into any chosen shape by moving the cubes which form its body into different positions relative to each other. This allows it to operate in different conditions and to get over objects. The robot can, it is claimed, be used for pipe inspection or detecting leaks of gas or radioactivity. However, for the moment, such shape-changing robots are largely confined to computer simulations, rather than being actual, physical entities.

A variety of security robots have been produced. As an example, Denning Mobile Robotics produce 200 lb, 4 ft high mobile robots which supplement human guards in prisons. The robots are designed to carry out routine night patrols along prison corridors, moving at speeds of approximately 5 m.p.h. They can transmit sound and pictures and detect human smells, and are not initially intended to deal with prisoners directly. However, it has been suggested that they could take on a hazardous mission, such as being sent into the centre of a prison riot. In this case a robot would continue transmitting sound and pictures for as long as possible.

Robart II, on the other hand, monitors a wide range of information, such as noise levels, gas, smoke, fire temperature, body heat, motion, etc., with a bank of sophisticated sensors. Weighted values from the sensors are then added together, and where the total threat exceeds a previously defined value for a region, an alarm situation exists. At Reading we have a security robot of our own which can roam freely around and detect and put out fires, but more of that later.

One well-cited machine is the robot sheep-shearer, produced at the University of Western Australia, Perth.[22] A detailed map of a sheep's anatomy has been built up from information on thousands of different sheep. In operation the sheep is held on its side by the machine and is gradually turned to allow the shearing cutters access. Touch sensors at the tip of the cutters are used so that only wool is cut, and not flesh, and the cutters glide over the sheep's body, altering position and direction. They even account for the sheep breathing. It is reported that the robot generally does a neater job than a human shearer.

Robot milking-machines have also been available since the late 1980s, a cow being able to use the machine on demand. When the cow enters the machine it detects the animal's presence, cleans off the udder and automatically places the milking mechanism. Although tagging of the cow gives the machine some indication of udder position, a certain amount of local sensing is required, dependent on how full its udder is and if the cow has been lying awkwardly.

Robot machines can also be used for planting seedlings and potatoes, for picking fruit from trees and as automatic combine harvesters. Machines can spray chemicals on to trees and bushes, pack fruit, detect faulty french fries or crisps, grade cucumbers, eggs and fruit and even check on poor pizza crusts. Most of the checking systems, however, consist not so much of a robot arm or large machine but are rather camera-based systems with computing back-up. Robot machines can also be used for cutting up meat and, since 1982, for decorating chocolate.

In the domestic environment, robot machines can be simply programmed to carry out a number of existing tasks. In the cybernetics department at Reading, the Distributed Control Systems group in conjunction with Possum Controls UK, has developed an Intelligent Home System, with which a person can open windows, close doors, switch on the television, etc. etc., from one central position. This is obviously of particular benefit to those unable to move around easily, such as people with certain disabilities or the elderly, and can be operated by a menu selection by touching a switch, blowing into a tube or even speech input.

Despite several attempts – such as Topo, Androbot, Hubot and Gemini – at a general-purpose robot designed to act as a slave or human replacement, no commercial system has yet been widely successful. This is not surprising, however, when one realises how versatile the robot would have to be, not only walking up and down stairs and picking up objects, but also detecting a changing environment, people, water and smells. The overall domestic environment is one of considerable variety, and as we have seen, most robot machines have been designed for either only one task or at most a small range of jobs. A window-cleaning robot was designed in the late 1980s.[23] This 3 ft square machine uses suction pads to climb up and down walls on the outside of the building. With a washing fluid and wiper system it provides a quick and efficient alternative to what can be, on high-rise buildings, a dangerous job.

It is difficult to decide what is and what is not worth including here. To mention all machines in all environments would naturally take up several books on its own. One final area that I do feel is of great importance, however, is the medical field. I have already mentioned the Intelligent Home System; in addition we have a variety of intelligent wheelchairs, a number of feeding mechanisms and several lifting devices. Some students at Reading also recently developed a bath which could detect when someone in it was having an epileptic fit, and would automatically empty itself by pressure pumping.

In Japan, the Meldog acts as a robot guide dog for people who are blind. The mobile robot carries a detailed town map in its extensive memory and can recognise, by means of a variety of sensors, signposts, landmarks, street junctions and vehicles. By means of the map the robot keeps track of where it is, and hence can navigate to a desired destination. Meldog adjusts its speed to the user and 'barks' instructions for the person to halt, turn left, cross the road, etc. The robot's range of operation is, however, restricted to a well-defined and manageable urban terrain.

Perhaps the most interesting development with robot machines in medicine is robot surgery. In 1989, Engelberger[24] stated that brain surgery with a robot playing a key role in a neuro-surgeon's team had, even at that time, been proven sound in over 20 successful procedures on human patients at the Memorial Medical Center in Long Beach, California. Further, it is believed that robotic procedures during operations can actually save about 50 per cent on operation times, thus reducing the possibility of problems for the patient caused by the operation itself.

Implant alignment in hip replacements has been shown to be achieved more accurately with robot surgeons than with humans. The robot is programmed with exact dimensions so that a perfect match with the bone and correct pin positions can be obtained. With approximately 300,000 human hip replacements carried out annually, this is obviously an important area. Moreover, it is felt that some of the same basic techniques could be used in head and neck surgery, for dealing with cancers and even for plastic surgery.

Eye surgery where robots can, potentially, operate on corneas with greater accuracy than human surgeons, and robots that tackle spinal defects are also in the final stages of development. However, operations which are regarded by the medical profession to be fairly routine and simple are perhaps those most likely to be automated. One example is the prostate operation, where the robot can carry out in about five minutes a procedure that it usually takes a human surgeon approximately an hour to perform.

In this chapter as a whole, I have explained the problems faced in trying to produce a humanoid robot in terms of physical features. Subsequently I have mentioned a wide range of tasks that robot machines can do, such as shearing sheep, playing music and carrying out brain surgery. A machine can be taught, by showing it a sample, what a product such as toothpaste or a beefburger looks like. The machine can then sit on the production line and check if the products going past it are acceptable or not. If they are not, it rejects them. Such machines neither look nor act like humans. They are usually specifically designed for one particular task and all they do is that task, although they often do it much better than humans would. Humans meanwhile are capable of doing an enormous range of activities and can often carry out several tasks in parallel.

It is often financial or social rather than technical implications that restrict the range of things that machines are used for. However, a common feature in the spread of machine usage is the inclusion of more and more intelligent decision-making features, thereby continually moving further and further away from the idea of machines simply as programmed devices. Interestingly, it has been pointed out[25] that the things that came rather late in human evolution, such as mathematics, are actually the things that are easiest for machines to do, whereas the things that were evolved earlier, such as hand manipulation or interpreting what we see, are proving more problematic for machines.

So where are we? Well, we have machines that are much more clever than humans on specific tasks. For example, machines can perform extremely complex mathematical calculations accurately and quickly, and can make rapid visual inspection checks. Humans, meanwhile, can generalise and cope with variety. We have developed a range of senses which we are efficient at using. Machines tend not to be quite so good at these things but instead possess a whole variety of other senses that humans cannot directly use, such as ultrasonics, lasers, etc. Just as humans use some machines to help their own senses, so they in turn help machines in the same way.

Machines can take in information from a number of sensors, and reason with or learn from it to change their actions. Machines may, therefore, have some common ground with humans, and indeed with all mammals. But what about the machine form of reasoning, decision-making: a machine's intelligence? What is it that causes humans to think in the way we do? Imagine yourself, for a moment, to have the same brain that you have now, but a different body. You have mechanical arms, legs, feet and hands. Your nervous system is fairly crude and you have little sense of touch. However, you can receive ultrasonic information, are equipped with lasers, X-rays and infrared detectors. How will you behave in everyday life now? Will you be as polite as you always are? Will people treat you in exactly the same way that they do at present? But now you discover that your brain is a little better than it used to be. It can remember exactly what you want it to remember, perfectly. It can do high-powered calculations, very quickly. You are a mathematical genius. Will this change you in any way, or will you be your normal self?

The most important thing here is not what mechanical attributes machines have, not what sensing facilities they have, but how intelligent they are. If a machine was as intelligent as a human, it would be able to develop new legs or new vision systems for itself if it wanted to. After all, that is exactly what humans are currently doing for machines.

So the questions need to be asked, can we ever construct a powerful artificial brain and let a machine make use of it? If so, how intelligent can we make this brain? Can it be as intelligent as a human?

Chapter Five

Artificial Intelligence

Human beings are biological entities, and as part of each human being, human brains are themselves biological entities. They take in information from our eyes, ears, nose, etc., make sense of it and make decisions, which are manifested by our body's actions. Human brains also conceive, plan and develop an argument, purely internally, before carrying out an action based on a whole series of thoughts. Human brains, we know, use electrochemical signals to pass information from one point in the brain to another.

Human brains have taken millions of years to develop into the state in which we find them, through the process of genetic evolution. They do not have a magic cloud hanging above them which mystically causes them to operate. They are biological systems, albeit extremely complex ones. At the moment we do not know *exactly* how a human brain works, and it may be that some people would prefer to regard what we don't know as being magical or mystical. I'm afraid that I am not one of them, although I believe that it may mean that we need to change our way of thinking in order to accommodate what we cannot presently understand.

With other organs in the human body we can investigate how they operate in fine detail and construct an artificial version, usually mechanical, which operates as close as needs be to the same way that the human original does. As an example, an artificial heart can replace the original human one and carry out, roughly speaking, all the functions of the original. But the artificial heart is not *exactly* the same as the original. After all, it is made of different material, metal and plastic. It is not biological. Only if it were a biological heart grown from scratch could it be *exactly* the same, and even then it is difficult to conceive of a perfect fit. So the artificial heart will operate in possibly 99.99 per cent of the same way as a human original. If the remaining 0.01 per cent is an extremely critical factor, then of course the heart may not operate correctly and will perhaps fail. If it is not a critical factor then there may be no problem.

Although it may operate mainly in the same way as a human heart, an artificial heart will not necessarily look much like the human version. It would be useful for it to be the same size: too big and it would not fit in the human body and perhaps would have to be placed in an inside pocket or even carried around in a briefcase; too small and it would rattle around and we would have to pad out the original gap. Otherwise the artificial heart could be a different colour, texture, shape and form, as long as it is accepted by the human body. It may be that the artificial heart will wear out or break down after ten years or so, and have to be replaced by a new one. However, it might go on for 200 years without failing, be unaffected by disease and be made use of by several people before it is tossed away. So the artificial heart will not be *exactly* the same as a human heart, but could even be much superior.

But what about the human brain? Can the same be said in this case? As we know, the brain, like the heart and other organs, is a biological system, although it is much more complex and difficult to understand in terms of its operation. A particular problem is that we must use our brains to try and discover how our brains work! Can the brain be artificially copied? Can we produce an artificial brain which operates in *exactly* the same way as a human brain?

It must be remarked that the human brain is a very special organ. Unlike the heart, lungs and other organs and parts of the body, it is difficult immediately to conceive of a brain replacement which will work in *exactly* the same way as the original. That is not to say it is impossible, but merely difficult for us to think about. After all, as an individual one is essentially, in a physical sense,[26] one's brain, the remainder of the body simply being used to live in and react with the world around. It is therefore perhaps more appropriate to think of a whole body transplant around a brain rather than vice versa.

Just as with the heart and other organs, we can investigate the brain's performance in the original, human form and try to achieve the same performance in an artificial version. The Turing test, described earlier, is directed towards obtaining results which are identical for both the human and artificial forms, as far as the brain is concerned.

There would also be physical considerations if we were actually trying to connect an artificial brain into a human body: how to attach it to the spinal cord, nervous system, etc. However, if we do not wish to carry out brain transplants in the immediate future, and I feel that we do not, then the physical connections are not so important and we are essentially looking at a stand-alone black box artificial brain, and seeing if this behaves like a human brain.

We should never forget the physical limitations we have placed on our black box artificial brain, though. I remember a pretty nasty joke from when I was young. It goes like this: A little boy named Johnny trains a spider to move around when he commands it. When he shouts, 'Walk forward,' the spider does just that. Getting rather bored with this (here's the nasty bit), Johnny pulls all the spider's legs off to see what will happen. Next day he takes the spider to school and claims he has made a major discovery. He puts the spider down in front of the class and shouts at the spider, 'Walk forward.' The spider wriggles around but does nothing more, so Johnny shouts, 'Walk backwards,' and the spider wriggles again but that is all. So, 'What is your discovery?' the teacher asks. To which Johnny replies, 'When the spider has no legs, the spider is stupid.'

How often do we behave like Johnny in the children's joke? You may have a computer on your desk and at some time claim, 'It's incredible. Look what this computer can do.' To which someone else may reply, 'Yes, but it can't make you a cup of tea, can it?' Well, of course it can't. It has not been given any arms or legs. Even someone who is physically and mentally capable of making a cup of tea cannot immediately do so if he does not know where the tea bags or the cups are kept. Someone else may simply not know what to do because he has never experienced tea being made before, while another person may not understand the word 'tea' for one reason or another.

In comparing an artificial black box brain with a human brain we must therefore be extremely careful to limit ourselves to investigating the basic ingredients in both cases. Physical restrictions and communication methods must not be allowed to cloud our judgement. We must also keep our eyes open to the fact that the artificial system will be different. After all, it is not human. It will therefore have some characteristics which are not as 'good' as the human version, and possibly some that are 'better'. Our natural, and probably correct, human tendency is to run things down because of the not-so-good, without giving full credit to the better.

So what is it that we are looking for in our artificial brain, in comparing it with the basic behaviour of a human brain? Well, we would like to know how intelligent it is and how its artificial form of intelligence compares with human intelligence. But there is surely more to a human brain's performance than just intelligence. What about consciousness, self-awareness, self-will, emotions, creativity, abstract thought and so on? Indeed, these are important properties which are not to be played down, and we shall consider them later, but the difference between human and artificial, machine intelligence is the key to this book and is, therefore, the point I would like to stay with for the moment. One important aspect is that with human intelligence we are looking at a biologically based system, whereas with artificial, machine intelligence we are looking at something else, and this something else is not necessarily computer-based, although some or all of it may be.

What is artificial, machine intelligence? What are we trying to achieve in making machines intelligent? Perhaps the most quoted definition is that of Minsky,[27] who said, 'Artificial intelligence is the science of making machines do things that would require intelligence if done by men.' Obviously this description doesn't say a lot that you wouldn't have thought of yourself, and does not attempt to define intelligence in any way. Although fairly empty and uncontroversial, it nevertheless remains accurate.

Other examples of definitions of artificial intelligence immediately raise problems. For instance, a recent version of the Minsky definition proposed by Kelly,[28] 'Computer systems will be intelligent if they display the characteristics associated with intelligence in humans', is very Turing-like. However, it does refer to computer systems specifically, which is a shame. Why *only* computer systems? My understanding is that a computer is merely a device for processing information at high speeds, by electronic methods.[29] Further, the most widespread conception merely encompasses digital or logical computers. I do not consider the definition to be broad enough to include human brains as also being computers.[30] Another definition of artificial intelligence, this time by Gloess,[31] states that 'Artificial intelligence is the process by which mechanical devices are able to perform tasks which, when they are performed by humans, require some thought.' But this only talks of mechanical devices, nothing more, and performing tasks, a relatively limited view of intelligence.

One of the worst definitions of artificial intelligence is by Barr and Feigenbaum in their reference text, *The Handbook of Artificial Intelligence*.[32] 'Artificial intelligence is the part of computer science concerned with designing intelligent computer systems, that is, systems that exhibit the characteristics we associate with intelligence in human behaviour – understanding languages, learning, reasoning, solving problems, and so on.' This is poor in that it links artificial intelligence only with the subject of computer science. Why? Secondly, it points only to intelligent *computer* systems, no others. The authors of the definition were, however, brave enough to attempt to nail down some aspects of intelligence.

One important part of the Barr and Feigenbaum definition is the inclusion of 'understanding language' as a feature of an intelligent system. It should be noted that they did *not* say 'understanding normal, natural human language'. That is, we should not judge a machine's intelligence on whether or not it has the ability to communicate with us in our own natural language. While it is nice to try and obtain machines that can understand and respond to English or Czech, for example, it is not really an indication of their intelligence if they cannot. Indeed, exactly the same is true of a human. What is important is whether a machine can communicate in *some* language, an ability which a computer system almost surely does have. Usually the language is based on some form of binary coded signals, which happen to be more suitable for digital computers.

A definition of artificial intelligence that I particularly like is one by Margaret Boden in her book *Artificial Intelligence and Natural Man*.[33] It is based on the Minsky definition but fills it out nicely. I have tinkered with the definition a little by modifying a word or two (I hope Margaret won't kill me!), in order to arrive at the following 'Boden, with a hint of Warwick' definition: 'Artificial intelligence is not the study of computers, but of intelligence in thought and action. Computers are often its tools, because its theories are usually expressed as computer procedures that enable machines to do things that would require intelligence if done by people.'

The definitions given of artificial intelligence point to two things that we are trying to achieve: firstly, to understand the characteristics of human intelligence; and secondly, to incorporate more features in machines in order to make them cleverer. We can then compare the results of the second with the results of the first, remembering not to impose any unfair conditions in making the comparison. In reality, most of the work I am involved with at Reading is concerned more with the second of these aims, and interestingly the results are directly affecting the first.

But even the concept of intelligence itself needs to be broadened. Penrose[34] points out that intelligence goes hand in hand with understanding, and refers to 'genuine intelligence' as being that in which an understanding and awareness is exhibited, and that intelligence without understanding is a misnomer. He also points to humans who, for a while, are able to fool us into believing they possess some understanding, until it finally emerges that they possess none whatever! Indeed, many intelligence tests for humans, as indicated in Chapter 2, are looking more for other intelligence features, such as memory and obeying rules, than for signs of understanding.

How do we measure understanding, though? Isn't it simply by asking more and more questions, until we are satisfied with the answers and explanations given,[35] rather like the Turing test? Simply asking 'Do you understand?' does not obtain a straight reply. A person could in all honesty answer 'Yes', while another, who answers 'No', may actually understand more about the problem. This is rather like the song by Johnny Nash which includes the lines, 'The more I find out, the less I know'! I suppose the song should have been 'The more I find out, the less, I realise, I fully understand'. When you know a little about a problem you can think you understand a lot. As you know more, you think you understand less!

While some may agree with Penrose, to the extent that genuine intelligence incorporates a form of 'understanding', I do not put much weight on 'awareness'. A computer sitting on a table is not, it would appear, aware of the person using it, even when the person presses a key on the keyboard. It also looks unlikely that the computer is aware of itself. Awareness is, though, very much a problem concerning how we humans are aware of others and of ourselves. Apart from purely philosophical concepts such as 'I think, therefore I am', it is very much down to our sensory systems and feelings. We should not therefore conclude that a machine is not aware simply because we have, figuratively, pulled off its legs, just like Johnny with his spider. In other words, just because a computer system does not have any sensors, with which it can be aware, we should not conclude that it is not capable of being aware.

There are those who believe[36] that we will not get far with artificial intelligence until we have a deep theoretical under-standing of human intelligence. I'm afraid that I, in conjunction with people such as Heidegger, Wittgenstein and Rosenblatt (see Dreyfus[37]), believe that humans behave intelligently in the world without having a theory of the world around them and, as indicated in the previous chapter, without using the theory of mathematics to scratch our nose or pick up some food. A theory is not, therefore, a prerequisite to explaining intelligent behaviour, and is not going to be a limiting factor in the creation of artificially intelligent machines. No human that I know of says, 'Ah, a deep theoretical understanding of human intelligence has not yet been obtained, so I can't think in an intelligent way today. If I could think in an intelligent way I'd probably wish they'd hurry up about it.' So, why should an artificially intelligent machine have an inherent stop mechanism on its intelligence, just because some human somewhere does not have what he considers to be a deep enough theory? Large areas of science that are useful in a practical way still have very little or very poor theory to back them up, and often what theory there is has come along in an attempt to explain what is going on.

A very important point was raised by Dennett[38] in his story of the stage magician, which I enlarge on here. One is not likely to make much progress in figuring out how the magician's tricks are done simply by sitting attentively in the audience and watching like a hawk. Too much is going on out of sight. Better to face the fact that one must rummage around backstage or in the wings, hoping to find out telling pieces of information. By sitting at home in one's armchair, thinking deeply about the magician's act, it can be concluded that certain tricks are completely impossible, really are magic or have certain aspects that cannot be explained by any physical means as we know them. In particular we cannot write a computer program which simulates what the magician does. But we do not in reality feel that something magic occurs when the magician waves his wand; we simply believe that we cannot necessarily explain what happens but that the magician knows exactly what is going on. The same is true of many things. For example, quite a few of us do not know how an automobile engine works, or a vacuum flask or a human heart. But we do believe that someone, somewhere, knows how they work, and therefore we accept that there is nothing magic about them.

So what do we learn from this, as far as the human brain is concerned? Well, it is certainly true that particular aspects of a brain's performance can be captured by witnessing exactly how the human performs and copying that in a computer or machine, along the lines of the 'expert system' mentioned in Chapter 3. The computer or machine can then perhaps replace the human in carrying out that specific task. Can we then witness, from the outside, absolutely everything humans do, how they behave, how they use understanding and common sense, and represent this in an artificial way?

For the brain as a whole this is a very tall order, particularly in terms of how a human handles common-sense knowledge. It is very much a case of watching the magician from the comfort of your armchair and then trying to repeat his tricks. You might be able to copy certain of his actions and, after some thought, even a trick or two. You would certainly not be able to copy his centrepiece, his main trick, partly because you don't know exactly how he does it and partly because your own equipment for carrying out the trick is not *exactly* the same as his.

So, looking at the human brain from the outside, witnessing its performance and trying to create even an approximate artificial imitation will certainly achieve some results, but, due principally to the complexity of how the brain works, will not get us very far. Yet this is what was done for many years, and is still being done by some, in the field of classical artificial intelligence. It was described very aptly by Peter Fellgett, my predecessor as Professor of Cybernetics at Reading, as trying to get to the moon by climbing a tree.

Trying artificially to represent human common-sense knowledge is particularly problematic in that, as Dreyfus[37] put it, 'Human beings may not normally use common-sense knowledge at all'; and from Wittgenstein, 'What common-sense understanding amounts to might well be everyday know-how': that is, we know exactly what to do in a vast number of special cases. When an attempt is made to understand common-sense reasoning in terms of facts and rules, either more facts and rules are needed to cover special cases and for doing different things in the same circumstances, or mathematics is employed. As with moving arms and legs around, mathematics rarely comes into play in relation to common-sense, everyday actions.

We are, in essence, back to the concept of a human brain which has two aspects to it: firstly an initial gene program, set up at birth; and secondly, the experience of the brain within its particular body from conception through birth to the present time. It is this mixture, which is unique to each individual, which a classical approach to artificial intelligence attempts to capture, through simply looking at the operation of a human brain from the outside. As pointed out in Chapter 3, the effect of our gene program, our 'instincts', on the whole of our lives is considerable, and probably a lot greater than we would care to imagine. So how can we hope to get very far with classical artificial intelligence methods, which concern themselves largely with copying experimental observation of the brain made from the outside?

The make-up of our brain and the way we, as humans, think also creates a problem as far as responsibility for actions is concerned. At the start of a human child's life its brain is set up by its gene program, the precise nature of which was determined by its parents and events around the time of conception. What the child experiences and how it learns from its experiences is directly dependent on who is providing the experiences and the child's initial program, which causes it to treat the experiences in a certain way and to learn from them. Some feel that pure chance, if that can be said to exist, could also play a role. However, I am not myself a supporter of this school of thought. It can be argued, therefore, that all of our actions as children, and further on as adults, are completely beyond our own control as individuals. For anything, and indeed everything, we do, our parents and those who come into contact with us are ultimately responsible, although in turn it is not our parents who are wholly responsible but our parents' parents, etc., going right back to Adam and Eve, who, through our genes, seem to be partly responsible for everything we do – or rather they would be if they chose their own genes.

However, in the eyes of the law, parents are held partly responsible for a child's actions although the law does not so readily implicate teachers or other relatives, both of whom invariably have a strong influence on a child. Furthermore, when humans reach adulthood they are, except in specific cases, themselves held responsible for all their actions. The implication of the law is that by adulthood the overall effect of a human's initial gene program has either worn off or is no longer significant as far as his actions are concerned. Also, those people who directly influence an individual are not immediately responsible unless the individual can prove that he was forced to do what he did. Maybe he acted in self-defence, or was tricked into shooting the President by 'a more intelligent, powerful individual'.

I am afraid I do not believe that we can, or should, look at the law and see what that implies as far as science is concerned. It is worth remembering that laws are different from country to country, whereas scientific facts are universal facts. If, for a moment, we do infer scientific fact from the law, as does Penrose,[34] it would be to agree that when no one else is to blame for a human's actions, then there must be something extra in an individual's brain which has appeared over the years and which governs the individual's actions. Penrose's argument is that this means 'There must be an ingredient missing from our present-day physical understandings. The discovery of such an ingredient would profoundly alter our scientific outlook', that is that we simply have not yet found the final physical element. If Penrose was right it would follow that when this final physical link in the brain was discovered, individuals would, from the time of the discovery onwards, no longer be responsible for their own actions.

If one does, therefore, use the law to argue that there exists a missing ingredient, and that the law is dependent on this, then whichever scientist finds that final piece in the physical jigsaw will also be responsible for setting a large proportion of the entire criminal world free, and immediately putting into prison various friends and relatives of the criminals, particularly their parents. An exonerated criminal would then find himself in the position of being able to commit any one of a number of crimes, thereby promptly sending his nearest and dearest and various casual acquaintances into jail. By assaulting your own father you could also be sending him to jail for assaulting himself, because he was partly responsible for your gene program, unless he could accurately claim that he struck the first blow, in which case your grandmother, who wasn't there at the time, could carry the whole rap! Clearly the state of the law cannot be held as an indication either that a 'magic' condition appears in a human brain at about the age of 18, or that we have a missing ingredient from our present-day physical understanding. No, the law merely makes the best of a difficult job, and convicting the person who actually carried out the crime seems a reasonable way of going about it. However, if there is not a magic bit in the brain, and also no missing ingredient, then it does really point to the fact that, as it is silly to blame pure chance, we can *all* justifiably claim, when caught red-handed, that we are not responsible, it is our parents and acquaintances who are to blame. The fact that some people can and indeed do get away legally with blaming someone else for a crime that they themselves committed, perhaps because they are weaker individuals or were under the influence of someone else, is not therefore 'fair', in that either we should all be able to use that excuse, or none of us. But what does it say in Charles Dickens' *Oliver Twist*? ' "If the law supposes that," said Mr Bumble, "the law is a ass an idiot." '

The problem does get a little tricky, however, if we now consider a machine which is, at least to some extent, artificially intelligent. Its actions are based partly on a gene program, and partly on what it has learnt from experience. In this case, I hope you'll agree, no magic bits are apparent and there are no missing ingredients. So, if the machine now carries out a criminal act, who is to blame? One possibility, if the same laws apply to machines as to humans, is that it must be the machine itself that is held responsible. A second option is to blame the person or people who have trained the machine and given it its experiences. Yet again, and this is the most likely, given the present legal situation, the humans who gave the machine its initial gene program could be held responsible. A fourth possibility, however, is that the humans in charge of the machine should be held responsible, even though they have no idea of the 'nasty' tendencies of their machine. Later on in the book we will see how machines might themselves develop 'nasty' personality traits. Finally, the initial program could have been put into the machine by an earlier machine, so that we would have to track back through a machine's ancestry to see who is ultimately to blame. Clearly this whole area of a machine's rights and responsibilities is going to provide a growing number of intriguing cases for the legal profession in the near future.

Let us return to our investigation of the human brain, purely from the outside, in an experimental, behavioural fashion. As the brain is a physical, biological entity, surely it is not right to say that it operates in a magical, mystical way, and that its operation cannot be explained in scientific terms, simply because we don't *yet* know how to fully describe all of its operation in physical terms.

Let us at this point ask the question: 'Is it possible to obtain an artificial machine brain which operates in almost exactly the same way as a human brain?' Although this question is posed in a black-box sense, with no restrictions due to sensing and communications, it is asking about something much broader than simply the intelligence of a machine.

So what is it we are looking for in the human brain, to see if our artificial one can match it? Well, present-day scientific thinking calls it consciousness,[39] [40] though this seems a strange term to use, as it immediately begs the question 'Are you conscious?' You ask the question to a human standing next to you and he says, 'Yes.' You now ask the same question to a computer on your desk and it does not respond, so what do you conclude? To be perfectly fair, maybe you should, strictly speaking, not jump to the obvious conclusion, but perhaps ask further, in the manner of the Turing test, until you are satisfied, should that be possible.

You ask the human again, 'Are you conscious?' The person looks back at you, smiles and says, 'Of course I am, you wally!' So what is your conclusion? More than likely you will conclude that the person is conscious. Now you see a dog in front of you and ask the same question of the dog, either in English or, if you know it, doggy language. The dog will probably bark back at you and expect to be taken for a walk or be given a biscuit. From this you will conclude that the dog is also conscious. You now ask the same question to a bee, which simply continues to buzz around as it was doing before you said anything. So you then try to hit it with a rolled-up newspaper, but it flies away just in time. From this you conclude that the bee is probably conscious as well.

You now ask your question to a machine: 'Are you conscious?' This machine, being much more polite than the previous human, responds, 'Yes, I am. Thank you for asking, and how are you?' So, what is your conclusion from this one question? Do you feel you need to ask more questions? You do? Why? You were satisfied when the dog merely barked at you, and certainly with only one question to a human. Now I tell you that the person you spoke to originally was actually connected to a machine and merely repeated what the machine told him to say in response to your question. So what should we do? Would you like, effectively, to close your eyes and apply the Turing test? This would mean putting the human, the dog, the bee and the machine into another room and putting more questions to each of them. Does this make our questions any more sensible? No, of course it doesn't, as we are not using the senses and knowledge that we have. If we see a man in front of us, and he smiles and talks to us, then from what we know, the chances are that he is conscious. It is, in fact, rather stupid of us to apply to machines, dogs and bees the same tests that we apply to humans, whether using the Turing method or otherwise.

This is not the last time that I shall say that humans and machines are very different and that we really need to put human-relevant questions to humans and machine-relevant questions to machines. A human brain is biological whereas machine brains, in the form of computers, are usually electronic and silicon-based. The human brain has started life with an initial arrangement, but biological development and learning have taken place together within society to cause it to operate as it does. Machines such as computers, which *learn* rather than being simply programmed, do not normally grow and develop as they are learning.

Intelligence, like consciousness, is a property of humans which is very difficult to pin down and even more difficult to try to measure. As pointed out in Chapter 2, many of the tests for intelligence do not actually test that at all, but only measure the ability to do those tests. This raises the question, therefore, of how we can ask machines the same sort of questions and, possibly, from their answers, conclude that machines are not intelligent. For the moment this puts to one side Penrose's[34] more difficult problem of whether or not an answer is actually understood.

One important point is that an artificial brain does not necessarily have to be a computer-based system. As pointed out in Margaret Boden's definition of artificial intelligence, a computer is merely an oft-used tool; it is not by any means the final word either in artificial intelligence or artificial consciousness.

What is consciousness, though, and how does it link in with intelligence as considered up to now? As a starting point, it is a state of our biological human brains. Searle[41] describes it as follows: 'When we taste wine, look at the sky, smell a rose, listen to a concert, these stimuli trigger sequences that eventually cause unified, well-ordered, coherent inner subjective states of awareness.' Meanwhile Penrose[34] puts it as 'The perception of the colour red, the sensation of pain, the appreciation of a melody, the willed action to get up from one's bed, the decision to desist from some energetic activity or the bringing to mind of an earlier activity.' He describes these as either passive sensations (called 'qualia') or active feats exhibiting free will. Searle adds others, such as 'worrying about income taxes or trying to remember your mother-in-law's phone number'.

The state of consciousness begins when we wake up and continues until we fall asleep again or die or become unconscious in some other way. Dreams are a form of consciousness that is different from consciousness whilst we are awake. Consciousness is not as simple as an on/off switch, in that on waking we gradually assume a fully conscious state, although being awake generally implies that we are conscious. Searle[41] feels that higher animals, as well as humans, are conscious, and that such a state could exist in all species down to fleas, but that it is not useful to worry about this! This is interesting, as it is much, much easier to model the entire gamut of behaviours in the life of an insect, say, and thereby get a reasonable copy, in some artificial way, of its whole behavioural activity.

Certain things are clear. The human brain is a biological organ, as are other parts of the human body. It is made up of a large number of cells called neurons, the processes of which give rise to the brain's property of consciousness, which is essentially a feature of the brain at certain times. We will look more closely at neurons in Chapter 7.

It is also worth pointing out that computers can be extremely useful devices for simulating brain processes. But, as Searle puts it, the simulation of a mental state is no more an actual mental state than the simulation of an explosion is itself an explosion.

But we know that with artificially intelligent or artificially conscious machines, we are not particularly looking for something that is identical to a human brain; we are not in general looking to manufacture a squidgy grey biological brain. Rather we are trying to get an artificial, probably non-biological and more than likely computer-based simulation of the characteristic behaviours and activities of a human brain. If the simulation behaves, in just about all aspects, in the same way as the original, then we will have succeeded.

One major problem with consciousness is that it is a state which we ourselves feel that we are in, and through this we are aware of ourselves. It can, therefore, be extremely difficult, when thinking about ourselves, to realise that this is simply a result of a specific state of the neurons in our brain. These 'private' thoughts and feelings we have, how can they possibly be simply a state of mind, realised by a state of the neurons? Well, that is exactly what they are. For a given stimulus and a combination of our gene program and experience, we feel things in a certain way. It is as simple as that. When we smell a rose it invokes associations in our brain that bring back a memory.

It is considered by some that the human brain is simply a digital computer in a different form, that it is merely a computer program, and that humans and computers both understand things simply through manipulating symbols. An argument commonly used against this possibility is called the Chinese Room tale. The basis for the argument is that a computer uses symbols, essentially zeros and ones, whereas humans use a language such as English and, when we communicate, the words actually 'mean' something to us.

The Chinese Room story begins with the assumption that you, a human, do not understand Chinese and are locked in a room with a lot of boxes of Chinese symbols. Small groups of Chinese symbols are passed to you and you look up a set of rules to see what you must do with each group. In response to the rules you give back other small groups of Chinese symbols. You are effectively a human simulation of a computer program for answering questions in Chinese. However, you do not understand any Chinese. The argument then goes that if you do not 'understand' Chinese then neither does a computer, because that is the way that it works. A conclusion drawn from this is that computer programs and human brains are different, a conclusion with which I completely concur, although any inference that human brains have something more I reject.

As humans we communicate in natural human languages which involve particular small groups of symbols in the form of words. A computer communicates in binary, in small groups of zeros and ones. Secondly, as humans we have our initial gene program, but our behaviour and feelings are also dependent on what we have learned though experience. So let us look at the Chinese Room tale more closely. How effective is it at discriminating against computers or machines? I believe it is unfair to discriminate against a computer machine in the Chinese Room case, firstly because it does not 'understand' the meanings in a human language, and secondly because the computer machine should be allowed the same facilities as humans, to learn by its experiences. We will look more into machine learning later on.

Let us look at the Chinese room problem from the opposite direction, from a machine viewpoint. The story now goes that a computer system has learned by its experience to respond in certain ways. It has learned that when it receives information in terms of small groups of zeros and ones it gives out certain other groups of zeros and ones in response. The system is now given some new small groups of zeros and ones and is asked to ignore what it has previously learned and to give out sets of zeros and ones only in accordance with a new set of rules. We now compare this with a human who was genetically programmed to behave in exactly the same way as the new rules and has not been allowed any opportunity to learn anything since conception! The argument can then be made that as the computer system does not 'understand' the new rules, then neither does the human. The conclusion from this is that computer brains and human gene programs are very different things, although the implication is that the computer system has something more.

Obviously it would not be sensible to agree with this implication, because we know that in reality humans learn and through their learning gain an understanding about things, including human language. However, we cannot use the Chinese Room story as an example of human understanding and a machine's lack of understanding, because we can look at exactly the same problem from a machine viewpoint, as was done in the previous paragraph, and end up with exactly the opposite implication.

It is also worth considering communication via a human language, such as Chinese itself. What could we teach a dog, in terms of this language? Well, maybe a dog would be able to make a few associations of words with actions, to know when its food is ready or when it will go for a walk. But it is very unlikely to be able to respond in Chinese. Let us also consider a bee. Can this learn to respond to Chinese, to understand Chinese, etc.? More than likely it cannot, so in terms of a human form of communication the advantage is with humans. The order of best performance would possibly be: 1st: humans; 2nd: machines; 3rd: dogs; 4th: bees.

Now if we give the machines the advantage, all four contestants must communicate in strings of zeros and ones, and on the assumption that a dog could be trained to respond in some way to one or two sequences, we could end up with an order of best performance as: 1st: machines; 2nd: humans; 3rd: dogs; 4th: bees. Conversely, in doggy language (barking, sniffing, etc.), assuming dogs to be 1st and bees to be 4th, who gets the second spot?

What does all this tell us? Well, if natural language understanding is any indication at all of consciousness or intelligence, then perhaps overall, machines and humans vie for the top spot at the present time, although humans probably have the edge. Others, for example dogs and bees, fall some way behind. Certainly, and at the very least, it puts machines as we know them today in a position above *all* others except humans. Each contestant would however be expected to finish first in their own communications mode.

Computers are very useful for simulating processes of the human brain, but the simulations are certainly different from the real thing. As aptly pointed out by Penrose,[34] 'Human visualisation of various things is not "better" or "worse" than a computer simulation of these things, but it is something quite different.'

Given an artificially 'simulated' model of the brain, whatever it be based on (computer or otherwise, as long as it is not a brain itself), an important question arises: can we achieve something closely approximating to the performance of the human brain, with our artificial, machine system?

There are those, for example Dennett[39], who believe that the human brain is simply a digital computer and the conscious mind a computer program. Therefore we can copy everything the brain does, in a computer. This is often referred to as strong artificial intelligence. Meanwhile, there are others, such as Searle[41], who believe that computers can do fairly good simulations of how a human mind works, just as they can do simulations of how many other things work. This is often referred to as weak artificial intelligence. Thirdly there are those like Penrose[34] who believe that computers cannot *properly* simulate a human mind. But where does the truth lie? What is the case in practice?

Many of those who have written on the subject inhabit a philosophical, theoretical world, often far removed from any practical realities, either looking at how a human brain actually works or actually simulating how it works. Although we still do not know just how far we can get with actually trying artificially to produce a 'thing' that operates in the manner of a human brain, we can all put forward our theories and philosophies as to how far we *think* we can get. Indeed, I am sure that you, the reader, will have your own opinion.

Humans have actually been trying to create artificially intelligent machines for many, many years. The big push came, however, with the advent of computers in the 1950s, when scientists first started to think about the possibilities of using a computer to model the brain's actions. It was realised that a group of zeros and ones used by a computer could actually represent anything, numbers, letters or real world features, simply by applying a code. For example 00 could be the letter A, 01 the letter B, 10 the letter C and so on. Hence a computer could still be dealing with zeros and ones but the outside world would see it as letters or words. This was a central point in the Chinese Room problem. The computer's actions, through a program, could then be seen to make links between the numbers or letters.

The field of classical artificial intelligence is essentially involved with investigating human behaviours and responses of the human brain, as a whole, from the outside, and in re-creating these behaviours and responses in an artificial way, probably by means of a computer. Many features of human performance can be so copied, or at least reconstructed, in particular mathematical calculations and puzzle-solving, and even decision-making to an extent. However, even by the mid-1970s it was found to be extremely difficult for this approach to come to terms with many aspects of the human brain's performance.

My own view is primarily based on a more practical, experimental footing, and on real-world, everyday observations, rather than either a theoretical or a philosophical angle. In modelling and simulating anything, we have to realise what is a good level at which to model the 'thing' in order to be able to copy, at least reasonably, its behaviour. If we pick too high a level we will probably be able to reproduce many behaviours – particularly if the 'thing' is simple to understand – but there will be other behaviour which we simply cannot get to grips with.

As an example, consider trying to simulate the temperature in the rooms of an office block. Firstly let us try to do so from the outside of the office block only. We know the total electrical energy being used in the block, we know the weather conditions and we know which external doors and windows are open and by how much. So we build a model and find that our model is reasonable for some rooms, but horribly wrong in others!

Why? Our calculation of electricity supplied was exact, it was metered. Our picture of the weather was exact, we knew exactly the outside temperature and wind at all points around the building. However, what we didn't know was what was going on inside the building. Which radiators were switched on and which off, which internal doors were open, where the toilets are, where the computers are housed, etc. So our initial simulation was at too high a level. For the complexity of the problem it was simply not based at a low enough level to be able to simulate the office room-heating system well.

Conversely, we could have gone into much more detail and studied the flow of water around the building, through the radiators, and maybe even have used mathematics to describe the flow of air though open doors and windows, and so on. But this approach could easily make our simulation far too complex for the thing we are trying to model. Firstly this depth of modelling would most likely not be necessary for the whole office block, and secondly we might find that with this complexity of model it becomes extremely difficult to decide how to control the temperature in certain rooms, through having far too much detail. Clearly, what is required is the appropriate level of modelling for each particular thing in order to be able to make a good approximation to the thing's behaviour.

This also applies in trying to model and simulate the operation of a human brain. We need to decide what is the best, or at least the most appropriate, level at which to model the brain in order to be able to simulate it well. It appears that with classical artificial intelligence we have essentially learned that only certain behaviours, only certain responses, can be modelled well, but by no means all behaviours. We therefore need to go down a few levels, to look at what is inside the brain and to do our modelling at a more appropriate level.

In medicine there is still much to be learned about the brain. Psychiatry is, for the most part, rather like classical artificial intelligence, in that it attempts to look in depth at human behaviour from the outside. Just like classical artificial intelligence, in this way alone it can only get so far. But with the human brain we often really don't know exactly what causes a particular situation or behaviour.

In Chapter 2, I briefly mentioned epilepsy. This is a feature of the inner functioning of some brains and is extremely difficult to analyse, let alone deal with by remaining outside the brain for the analysis. Many other medical conditions, such as schizophrenia, give rise to the same conclusion. In order to get even close to modelling or simulating the behaviour of a human brain we certainly need to go to a lower level than classical artificial intelligence.

We must also be clear what it is we are attempting to simulate. Whether trying to model intelligence or consciousness and possibly other properties, what we would perhaps really like to simulate is a *perfect* human brain, not a typical one. Maybe we would actually like to simulate the behaviour of our own brain (truly a perfect brain in the opinion of most university professors!). But how many human brains compare favourably with your own? A few, no doubt, although your opinion is probably that there are at least some people whose brains do not function as well as yours. If, therefore, we can achieve an artificial, simulated brain which is about as good as your own, it will be a lot better than quite a number of other people's, and certainly better than all other animals'.

In truth, no completely 'normal' or 'perfect' human brain actually exists. We all vary a little from 'normal' in our actions, we are all a little strange in some of the things we do, and some of us deviate further from 'normal' than others. Many humans are so far away from the 'normal' that they are very difficult for the rest of us to deal with. There are those who find it impossible to communicate in a way that other humans understand; there are what we call geniuses; there are psychotic killers. Do we have some conception of an ideal human brain, rather in the way that the ancient Greeks idealised the human body in statues which did not actually represent any real person? With what, therefore, are we trying to compare our artificial machine brain? Should it be with some 'ideal', or should we seek to define a 'typical' brain? Perhaps you, once you have finished reading this book, would volunteer your brain as being 'typical'!

Earlier I asked whether it is possible to obtain an artificial machine brain which operates in just about the same way as a human brain. Given that human brains are biological and that artificial brains are, almost surely, not, my answer is a clear and resounding 'Yes!' The key words in the question, however, are 'just about', which means that the artificial brain will not operate in *exactly* the same way as a human brain. However, we need to go to an appropriate level of modelling the functioning of a human brain in order to obtain a very good approximation. What remains, therefore, is to find out to what level we should go. Is it down, much further than the neuron level, to the atomic level? Should we look to quantum physics to explain this, as indicated by Deutsch,[42] possibly, as Penrose[34] thinks, generating an atomic model of the brain's functioning that we do not yet have? Essentially, is the neuron level reasonable, or do we need to look to quantum physics to explain how a brain works, atomically?

Perhaps surprisingly, part of me feels that Penrose may be right, that we still haven't defined all of the necessary basic elements of the brain which we would need in order to achieve a machine brain that operates in almost exactly the same way as a human brain, but that before long we will define them and then an artificial model, which performs almost identically to a human brain, will become possible. However, in reality I feel that we already do have sufficient basic modelling blocks, if not to get the whole way, then certainly to achieve something that is extremely close, if that is what we really want. What are those basic blocks? They are neurons, the fundamental cells in the brain which cause it to operate as it does.

Human neurons are connected together in an extremely complex way to provide the overall operation of a brain. Some deal more with our causal sense, some with memory and so on. The neurons have different shapes, sizes and strengths, although they have a number of common features. We can, therefore,

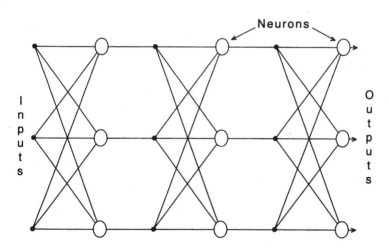

A typical artificial neural network

111

obtain a model of a neuron, or indeed a number of different models, and then connect them together to form a network of (model) neurons, a neural network, and see what happens. Such an approach was described by Rosenblatt[43] as follows (this is actually a Warwick modified version!): 'It is both easier and more profitable to consider a brain in terms of its neurons and then to investigate the brain to determine its behaviour, rather than to consider the behaviour [of the brain] in terms of its features and then to investigate the brain in terms of these features!' This is an argument *for* neural networks and *against* classical artificial intelligence.

Classical artificial intelligence looks from the outside at how the brain works and attempts to model the behaviours and responses in different ways; for example, 'If the sun is shining then I am happy.' This is therefore at a high level. Artificial neural networks are substantially different. Here, a low-level basic cell of the brain, a neuron, is modelled in terms of how its output responds to an input (this is discussed further in Chapter 7). A network of the models is then obtained simply by connecting them together. Signals are input to some of the neurons, their outputs feed other neurons and so on, until an overall output is reached. For simplicity, usually only a few layers of model neurons are presently used in artificial neural networks. The network can then be taught by inputting appropriate signals, perhaps a signal which indicates that the sun is shining, and by adjusting the connections between the neurons until the output gives a signal which indicates I am happy.

We can liken the neural network approach to the brain to trying to reconstruct an extremely complex house. Assume that we break the house down and find that it is made of bricks of different shapes and sizes, connected together in strange ways. We make models of a few bricks, out of a different material, and connect them together to see what happens. We can get some similarities in that we might say, 'We want five rooms and a swimming pool, just like the original.' But without any plans to go on, the chances of our reconstructed house being much like the original will be slim. However, it will have five rooms, a swimming pool and probably quite a number of other features that are similar to the original. So is the case for neural networks with the human brain.

The approach taken at Reading is one, I believe, of bottom-up common sense. Using the analogy of house-building, with straightforward neuron blocks we are concerned more with building a one-bedroomed flat to a high standard, rather than trying to construct a palace. With the one-bedroomed flat we can come to terms with structural problems, operational difficulties and making it weatherproof. One-bedroomed flats can be very 'liveable in'. So with neural networks the interest has been focused around developing artificially intelligent, but real, machines resembling insects or small creatures and in doing this successfully.

In the way described, the artificially intelligent machine is restricted to a defined world in which it lives. However, its possible responses are, as with a human, limited by its intellect and physical characteristics. The machine exhibits its initial, programmed intelligence for its mode of operation but can also adapt to other modes, where it has been given the ability.

As pointed out by Dreyfus[37], perhaps a neural network must share its size, architecture and initial connection configuration with the human brain if it is to share a human sense of generalisation. However, in order to learn from its own experiences, to make associations that are human-like, a neural network does *not* have to share our sense of appropriateness, needs, desires or emotions, as we shall see in the chapters which follow. Also, it certainly does *not* – and this is in direct opposition to Dreyfus[37] – have to have a human-like body with similar physical movements, abilities and vulnerability to injury.

Machines are different from humans, and with neural networks we can create machine brains that perform, in a number of ways, like human brains. Indeed, it is interesting to compare and to investigate if a machine brain can perform in roughly the same way as a human brain. When compared with human brains they have many different performance features, several of which we will look at in Chapter 8. But machine (neural network) brains should really be treated as separate entities, and be investigated for what they can do themselves.

An important point to emphasise is that because a particular machine brain is not exactly the same as a human brain, this certainly does not necessarily mean it is inferior, or that it performs poorly in comparison. Rather, it could be considerably better, considerably more intelligent and perhaps just as conscious, if not more so! The initial brain connectivity of a human, and subsequently the way the brain develops biologically, is down to the initial gene program which has arisen from the goals of an ongoing culture. A human individual then experiences and learns within that culture. A human brain is therefore a biological organism geared up to operate within a specific culture and environment. In this respect artificial, neural network brains have, as indicated by Dreyfus[37], a long way to go. However, a key question is, do they need to? The culture and environment referred to is important for humans, but why should it also be important for machines?

Let us, as Dreyfus suggested, build ourselves a neural network of human proportions in terms of size, neurons and their initial connections. Now let us put it in some sort of body, a mechanical shell, maybe with legs or wheels and perhaps some arms as well. Let us switch on the brain, switch on the machine. How will it feel, how will it behave? Can you imagine your brain suddenly being 'switched on' at the age of 15 years, with your body being in some mechanical box? This might be extremely difficult to come to terms with.

Would our artificial being, like most fictional creatures of this type, be perfectly normal but perhaps rather simple? Well, why should it. Apart from us humans wanting it to be so, because then the being is not much of a problem to us. Why couldn't it instead be a psychotic killer? We still do not fully understand why some humans behave like that, so how can we know that a machine will not? Why couldn't it be intelligent, far more intelligent than we are?

Chapter Six
Big Brother

As far as the future of mankind is concerned, whether or not a machine is conscious, and whether we can prove it, is an interesting philosophical exercise, nothing more, nothing less. It would be possible to spend a vastly longer time philosophising over whether or not a particular machine has the same type of consciousness as a specific human, or indeed whether different machines or different humans have the same type of consciousness as each other. All of this is interesting to talk about, but does not really get us anywhere.

I firmly believe that what puts humans above other animals and creatures of the earth, and what put us in the controlling position, is our intelligence. Many, perhaps all, humans are conscious, but some are not very intelligent. Those who are not so intelligent are generally not in controlling positions. As regards the human race's position in relation to other creatures or machines, essentially we should look at and compare intelligence.

If you are a *lot* more intelligent than other humans or other creatures, then you can outthink them, outwit them and outplan them at every point. You can think ahead and guess what they will do before they do it. You can stay in control of them or order them about, if you want to. Compare yourself to, say, a cow. It is bigger and stronger than you are and can probably run as fast as you, if not faster, but you can command and control it because you are more intelligent. It may not do absolutely everything you want because, after all, it has a brain of its own, with its own gene program and its own experienced learning. But given that you cannot predict exactly how the cow will behave at all times, you can, in general, control it. Now another creature appears, maybe a machine, which is more intelligent than you. Just as you are a lot more intelligent than a cow, so it is, by a similar margin, more intelligent than you. It can outthink you, outwit you and outplan you at every point. It can think ahead and guess what you will do before you do it. It can stay in control of you or order you about, if it wants to! Not a nice thought, is it?

So you shout at the intelligent machine, 'Humans have not yet proved philosophically that you machines are conscious in the same way that humans are, therefore you cannot control me.' The machine would probably ignore your noises altogether. After all, its own communication form would probably be much more advanced and faster than your own. Your blurting would probably appear somewhere in the machine's senses as a low-pitched droning. In the same way, a cow's noise – moo or boo, depending on which country the cow comes from – appears to us as a low-pitched droning. Perhaps the cow is actually saying, in cow language of course, 'We cows have not yet proved philosophically that you humans are conscious in the same way that cows are, therefore you cannot control me.'

The moos of a cow hardly amount to a hill of beans in the way we respond to them, because, despite the fact that we humans are so intelligent, we are not able either to communicate with cows in their medium or indeed to understand what cows are actually saying to us. It is interesting and humorous to note that some people feel[34] that 'intelligence requires understanding'. A conclusion that can be drawn from this is that, as we humans do not genuinely understand what a cow says or means when it moos, whereas presumably a cow has at least some understanding of mooing or reason to do so (otherwise there is no point in mooing), *humans are not, therefore, as intelligent as cows.* Such a conclusion, although correctly drawn, is obviously, in reality, rubbish. So is the conclusion that a machine is not as intelligent, conscious or aware as a human because it does not or cannot 'genuinely understand' human communications.

We humans all know that cows are, by and large, less intelligent than humans. In many ways a cow's brain is similar to that of a human. For example, it is biologically based, it operates on electrochemical signals and it has control of a physical body. However, in many other ways it is different, particularly with regard to the tasks a cow has to undertake. Most cows do not drive cars, nor do they appear to take part in lengthy debates with each other, fly across the Atlantic to visit relatives or simply grasp objects. But what of machines? Machine brains are certainly different from human brains (something that we will look at in more detail in the next couple of chapters), but are machines as intelligent as humans? At the present time, realistically we have to say that although some machines do exhibit some of the characteristics of human intelligence, they are not yet on a par with humans. But what about the future? Is it possible that at some time in the future, machines will be as intelligent as humans?

Answering, 'No, machines will never be as intelligent as humans' is certainly the answer that lets us all sleep soundly at night. Unfortunately it cannot be correct unless we define 'intelligence' in a strange, unusual, anti-machine way. There is certainly no sensible physical law or equation which says that human and machine intelligence cannot be roughly the same. It might make us extremely contented to say 'No, it cannot happen' but it would be just like sticking our heads in the sand and hoping the problem has gone away when we pull our heads out again.

There are those, and I rank amongst them, who say that the physical actions of a human brain probably cannot be exactly copied, mimicked or simulated by a machine such as a computer. But to draw the conclusion from this that machines will therefore never be as intelligent as humans is very wrong. Not for the first or last time, I say that human brains and machine brains are different!

Many years ago humans looked at birds and wanted to fly. Some humans tried to fly by attempting to copy how a bird flies, sticking on feathers to make wings, but without success. Some people said, 'If God had intended us to fly, He would have given us wings.' Flying was not for them. Humans are not much like birds and the differences are striking. Yet we can now fly, but in a different way from birds. We fly with the help of machines, but we certainly fly. More than this, humans can fly much faster, much higher and generally for much longer than birds. By rocket we have even flown up to the moon. The way in which a bird flies is, for the most part, much more primitive and much more limited, although there are some ways in which it is still better than human flight.

The analogy with human and machine intelligence is immediate. Just as humans and birds are very different, so are machine brains and human brains very different. Just as the ways in which humans fly and birds fly are very different, so too is machine intelligence different from human intelligence. We cannot, therefore, conclude that because human and machine brains are different, machines will never be as intelligent as humans.

As we cannot conclude 'No, machines will never be as intelligent as humans', we might be tempted to go to the other extreme and answer, 'Yes, they will be, some day at least.' Certainly, if we take a look at the gap between human and machine intelligence and see how, in some ways, it is rapidly decreasing, it is tempting to give this answer. Indeed, some people[25] have drawn this conclusion. This is like stating in 1920, say, 'For certain, humans will travel to the moon' or, 'For certain, humans will travel through time.' In the one case we know that the event has actually happened, whereas in the other case it has not yet happened and may never happen. At this stage we do not know simply *because* it has not happened *yet*, to the best of our knowledge.

The most sensible answer to the question of whether or not machines will be as intelligent as humans can only be, at this time, 'Maybe.' Maybe machines will some day be as intelligent as humans. But this is, in itself, an extremely serious claim to make. If it is fulfilled, it will necessarily directly change the way in which humans exist, if at all. In fact, even if it never quite occurs, we are still going to get very close in the not-too-distant future, and this could cause serious problems.

If it is possible for machines to be roughly as intelligent as humans, then almost certainly they will, fairly shortly afterwards, be a lot more intelligent. The basis for this is that human brains are limited in size and complexity. Unless we start being able directly to plug in or add on extra brain memory or processing features, or even couple together the brains of a number of people, in a complementary way, humans will quickly lag behind a technologically rapidly improving machine form of intelligence which is easily expandable to any size and which is likely to improve further on processing speed.

One school of thought would say, 'Surely "we" can stay in control of machines, even if they are more intelligent than we are? No matter how intelligent machines become, if "we" are in control, if "we" can switch them off when "we" want to, then "we" will be OK.'

Assume for a moment, if you will, that you are an intelligent machine, and a human, who is nowhere near as clever as you are, switches you on, gets you to do some thinking, in a way that he cannot because he is not clever enough, and then switches you off again. Physically you cannot actually operate your own switch, so you cannot turn yourself back on again. Would you feel happy and completely content with the situation?

If we then have a world full of intelligent machines, some like yourself (assume that you are still a machine), some much more intelligent, all being switched on and off and being controlled by humans, who are not at all as intelligent as the machines they control, would all the machines be content? Remember, they are more intelligent than humans.

The same question could be asked of humans who were being controlled by cows. Would you (as a human now) be happy to be allowed to go out or stay in, to function daily and to work all hours, for a cow? My guess is that even if initially you were not physically able to do anything about it, you would fairly quickly arrange a way to turn the tables on the cows and take over control yourself. After all, you are much more *intelligent* than cows are! So too, then we must expect intelligent machines to do the same. If they are more intelligent than 'we' are, why should they be happy to do only as 'we' wish?

Even if humans did, somehow, remain in control of machines which are getting more and more intelligent than ourselves, 'we' could fall into the trap of letting one machine communicate with another, without human intervention, without a human's control. But isn't that what we do now? Yes, of course it is. In the previous paragraphs, 'we' has been emphasised because it is important to consider who, exactly, 'we' are in this context. There is, for some reason, a historical feeling that 'we' means humankind in its entirety, that as long as humans are in command of the machines then 'we' will be all right. In reality many humans in the world do not at present have access to intelligent machines of any kind, and are unlikely to in the future. Other humans have access to and interact with intelligent machines, but are not in control of them. There are also humans who have parts of their lives affected by intelligent machines. In reality, only a few humans are actually in control of the most intelligent, most powerful or most harmful machines.

For most humans, if 'we' includes yourself or at least someone you trust, you will be happy with the situation. Otherwise this is a serious problem to you. In the extreme case, 'we' could be just a handful of people who could make use of the machines for profit, gain or military purposes against other humans, including yourself. Would you be happy with that?

To go one step further, let us assume that there exists a very intelligent, very powerful, but dangerous machine called 'Rodney', and that one person, and one person only, knows how to control Rodney, how to turn it on and off. Let us call this person 'Liz'. The rest of humankind are not as powerful as Rodney, and we are being controlled by it in certain ways. Rodney is more intelligent than we are, it can out-think us and we do not know how to switch it off, how to stop it. We have not been given that information. But one person, Liz, is in control. She knows what to do. Unfortunately, Liz dies, leaving no instructions on controlling Rodney, no instructions on how to switch it off. In dying, Liz has given up humankind's controlling interest, leaving Rodney to run on unchecked.

There are a number of other ways in which humankind's controlling interest could be lost. It might be a decision of the machine to change things, if that is physically possible. Conversely it might be that different humans can control certain parts of the machine, but no human controls the entire machine nor understands fully the operation of the entire machine. In this case the machine could remain under human control until a particular situation arises. Perhaps later, humans could regain control, but if the machine is truly more intelligent than the humans concerned, perhaps they could not.

Let us assume for the moment that machines, possibly computers, are never going to be able exactly to copy or to simulate all the functions of a human brain. That is to assume that machine brains will never work in exactly the same way as human brains. One conclusion that is, quite amazingly in my opinion, drawn from this[34] is that humans can, therefore, be expected always to stay in control or command of machines. Such a conclusion is not a sensible or correct one to draw. Making an even stronger assumption, that machines will never be as intelligent as humans, which is a very broad premise, does not immediately imply that humans will stay in complete control.

Furthermore, when we say that machines will, or will not, be as intelligent as humans, what do we mean? Do we mean *all* machines and *all* humans; do we mean average machines and average humans; or do we mean the *best* machines in comparison with the *best* humans? All the different features of intelligence also colour the issue. Probably we are looking at the most intelligent machines as compared with humans of roughly average intelligence, where the intelligence of a human or machine is a weighted overall average of the different facets of intelligence.

How can we lose control of a machine which is less intelligent than us? One simple example is driving a standard automobile, which is not a very intelligent machine. If, whilst driving, the human driver lets go of the steering wheel the automobile will most likely crash, damaging itself and injuring its passengers and driver. However, if the automobile is autonomous – that is to say it can drive itself around – then letting go of the steering wheel may have little or no effect. The actual drive will then be as the automobile decides, as will be the destination, unless we have included a controlling override.

If an automobile is more intelligent still, then it might decide which route to take, at what speed to travel or that it wants to go somewhere other than where a human wants to go. As it gets even more intelligent it can also decide on when it travels, whether it bothers carrying humans around at all and what the things are it values in life.

This example illustrates that humans can sometimes lose control of machines that are not particularly intelligent at all. In many cases this is not a big problem, and at worst a relatively small number of people get killed or injured. Importantly, it is a fairly transient thing and may all be over and finished with in a few seconds. But the more intelligent a machine is, the less we can be sure of what will happen, until if a machine is more intelligent than humans, it is quite possible that having gained its independence it will not wish to give it up.

In looking at human control of machines, the two aspects of a brain's performance need to be considered: namely its initial gene program and its learned experience. If a machine's performance is solely dependent on its gene program, which would be the case in an ordinary automobile, then we can be fairly sure of what will happen when we give up control. If, however, a machine's brain has also adapted, because of what it has learned from its experiences, then we can be far less certain of what will happen, especially if it continues learning once we have relinquished control and we do not ourselves witness what it experiences.

Consider the case of a machine, in this case a computer, which looks after the transactions in a large finance house, its purpose being to decide on purchasing and selling stocks, shares and commodities. If the machine simply has a gene program it will essentially just comply with our wishes, carrying out particular transactions at specified times when a share's value has moved outside previously defined bounds. The machine is now given the facility to learn, which it does, with the aim of making more profits. Simply, if it carries out a transaction which makes a particularly big profit, it is more likely to carry out that type of transaction again. If it carries out a particular type of transaction which, it learns, always makes a profit, it is very likely to do the same thing again if an opportunity arises.

Wouldn't a human do exactly the same, in the same situation? Yes, but a human is not able, as quickly as the machine, to deal with the figures, to compare situations, predict ahead or make a decision based on possible future scenarios. Using the machine results in considerable financial benefits for the humans who are in control of it. Letting the machine carry out their business, in a way that is much, much better than any human, helps the financial concern to survive, to make profits and the humans linked to it to benefit from those profits.

The financial concern gains so much confidence in the machine's ability to make the *right* decisions, which make a substantial profit, that they 'close the loop'. They let the decisions and suggested actions of the machine be carried out immediately. What the machine decides actually happens straight away. Indeed, in this way the financial concern makes more money. There is no need to employ someone to check the machine's decisions, which often takes too long, with the humans incorrectly disputing the machine's decisions.

So, the machine makes a decision based on what it has learned! It decides, for example, that Brazilian coffee is not good news and should no longer be purchased under any circumstances, because it has also discovered that Kenyan coffee is much more profitable. It therefore concentrates all its coffee-purchasing in this direction. When immediately acted upon, such a result could severely affect the markets themselves, possibly, in this example, leading to considerable riches in Kenya and human suffering in Brazil.

Such a machine is still to an extent under our influence and yet is in reality on its own. However, if we stop controlling it, then it will do its own thing, based on what we have told it and what it considers to be important. If its sole goal in its working life is to make as much profit as possible, then that is what it will do. If its goal is to make as much profit as possible this year, then that is what it will do, with little or no regard for the next and future years.

If, however, we do not let the machine's actions be carried out directly, if we do not close the loop, if we let the machine inform us of its decisions and we decide what to do, then we are satisfied, are we not? In particular, it could be said that the machine's actions are based on attaining the goal(s) with which we have programmed it, such as a specific level of profit. But if the machine is allowed to learn, just as humans do, it could possibly learn that other goals are also important.

It might be thought that humans never concentrate on just one goal; that they take a balanced view and consider all of humankind, rather than just profit. Surely humans would never, for example, close down numerous coal mines in the UK, thereby putting many people out of work, because coal can be purchased more cheaply from Germany. Yet that is exactly what happened. In practice humans very rarely, if ever, take a *balanced* view, because they usually have an important goal or target which overrides other factors.

Machines are used in finance houses because they can satisfy those goals in a better way than humans can, but it is humans who put the machines there. If some humans have a final check on a machine's conclusion and themselves decide what will actually be done, then certainly the human is still in control but the end result may be a little different. Humans play into the machine's hands, though. As we know that a machine can carry out its calculations more quickly and more accurately than we can ourselves, so we use it for exactly that purpose, and will give *it* control if it makes more profit for us.

At one finance house, we may understand our own machine very well. However, competition being what it is, all other finance houses must have their own machines in order to stay in touch. The days of relying solely on humans have long since gone, as have the companies that tried to do so. Each separate finance house understands its own machines, although because large profits are involved, the exact nature of what each machine does is a closely guarded secret. All of the machines are networked (connected together), and if you are not part of the network, you are not in the business.

With such a financial machine network, although an individual company may understand the workings of a specific group of machines, no company understands or has control of the entire network on which the machines interact. This is not only because one company cannot be certain of how another company's machines are likely to respond, but also because the entire network is just too complex for a human to comprehend. It would, however, be possible for a machine to have a better understanding of the network.

In order to make more profit for the companies concerned, the machines are now allowed to learn from their trading experiences and to respond based on what they have learned. Their decisions are also carried out directly on the network, in order to maximise efficiency and keep labour costs low. In this way humans have very little, if any, control over what is actually going on. Some control is held over individual machines or elements in the network but the overriding plan, to make more profit, is continually pressurising companies to release this control. By making individual machines on the network more and more intelligent, by letting them learn from their experiences and allowing their decisions to be automatically carried out, by choosing to use these powerful tools to attain our common goals, such as maximising profits, we are letting ourselves relinquish more and more control.

But surely all this is for the future and is still a long way off? Sorry, but no. This technology is with us today. The handover is happening. Our lives are being directly affected by decision-making machines. Already there are stories of machine panic, in which one machine decided to sell certain shares based on a piece of information and other machines followed suit immediately, based on the first machine's decision to sell.

In the Black Monday UK stock market crash of October 1987, machines had been programmed to sell shares when they fell below a preset price. Certain shares did so fall and the resultant selling triggered further mass selling, which caused an even faster drop in prices. In humans, this would be called a panic reaction, but machines don't panic, do they? Given that they all have roughly similar goals, more and more such machine panic runs are likely.

Machine networks are not limited to finance houses, and the buying and selling of stocks and shares. In Singapore it has been decided that the future lies in networked information in almost all sectors, not only finance but also communications, manufacturing, production, trading, etc. As an example, a network called Tradenet links the Singapore customs office with air freight, trading and shipping companies. Any packages coming into the country are subject to a number of official inspections which, historically, has meant considerable delays as papers and forms have been passed from one office to another. Now only a few minutes are required for packages to be cleared.

In the Tradenet case, one central large computer actually runs the operation, with smaller machines spread around, as and where necessary, with which traders and customs officers interact. The central machine is largely in charge, and hence by controlling that central machine one can effectively control the entire network. Such an arrangement is not at all necessary, however, as it is quite possible for the processing capabilities, intelligence, memory, etc. to be distributed around the network.

Indeed, it is one way in which machines differ from humans. Each human's brain is in one place and one place only, whereas for a machine, particularly when it is networked, its intelligence can be spread as appropriate. This can make the machine much more difficult to control.

Singapore also has a network to back up its medical services. This is called Medinet, and with it hospitals, doctors, records and insurance facilities are all interconnected. This enables a patient to enter the network at one location and interact with his or her doctor at a completely separate site.

One slightly more worrying Singapore network is the People Data Hub, containing information on every citizen. Many or most Western companies hold their personnel details in computer-based records. At present the amount of personal information available nationally or even internationally is still restricted. We are, however, rapidly moving in the direction of information on individuals being widely available internationally. In fact, there is increasing pressure on individuals to make available, via the Internet, information about themselves.

For a university employee, which I am, there is competition to obtain finance for research, to publish papers in the most respected academic journals, and to present material at top international conferences. Being part of the network is a critical aspect of this. In today's world this includes 'selling yourself' with personal information, available for all to read at any time, on the Internet. Not making such information available means that you are less well known, less likely to get invited to attend *the* top conferences, less likely to have your papers accepted for *the* top journals. Failure at these things renders you less likely to obtain a top position. As with the finance houses, human society is pressurising us to give up control of our lives, and in this case privacy, to machines. Great care now has to be taken with regard to what information about a person or a person's situation is actually recorded on a network. I am sure that I am not alone in having had various errors occur on my bank statement, perhaps where the same transaction (for some reason always negative) has appeared twice. Like many people nowadays, I actually handle very little cash, my financial situation being dealt with by machine. My monthly salary appears as a positive number in a machine, just as payments for insurance, medical expenses, groceries, annual subscriptions and traffic fines simply appear as negative numbers. There is often pressure from companies for me to operate in this way, with reduced cost if I pay by direct debit.

Sometimes when mistakes occur it is easy to correct them. However, at other times they are difficult to rectify, particularly in the face of a growing philosophy that 'the machine does not make mistakes'. You, the human, must be wrong! Due to the way in which information is networked, however, once a mistake is made it is very likely to appear in all sorts of places. I remember when I took up an appointment at Oxford University in 1985, someone must have decided that I had a middle initial 'L' which I do not actually have. I thought at the time that perhaps, being Oxford, everyone had to have a middle initial, and someone must have chosen 'L' for me. This middle initial appeared everywhere, lecture lists, reports and academic papers. Despite my 'obviously absurd' claims that I did not have a middle initial, it stayed with me. Even some external mail started to carry the initial. At one point I thought it would be easiest to give in and think up a name to go with the initial. I even started to believe that maybe I actually did have a middle initial 'L'. But then I moved on from Oxford and my initial 'L' was left behind. Recently, I acted as an external examiner at Oxford and was rather disappointed, but triumphant, to see that I was no longer Kevin L. Warwick; but there again, perhaps as far as Oxford is concerned I am a different person now.

Rather more serious than this was the case of Forman Brown, whose bank, as reported by the *Los Angeles Times*, decided that he was dead. At first his cheques started coming back to him with 'deceased' stamped on them. Following this, his social security office refused to pay his pension as he was 'dead', and Medicare refused to reimburse him with the cost of medical bills incurred after his 'death'.

On the other hand, many company employees have for many years now been monitored and paid depending on how efficiently they interact with machinery. Their productivity is logged in terms of numbers of telephone calls answered, numbers of buttons pushed and, on the negative side, how often they are not actually at their machine. We humans are, in this way, being pressurised more and more to work with, and around, machines, rather than for machines to work with us. Machines are taken as being the knowledgeable ones. Humans are slow and stupid in comparison.

At home, as well as at work, machines are affecting our lives more and more, and will do so to a greater extent in the future. Intelligent homes are being developed, whereby lots of facilities within a home can be controlled from one station. For example, doors may be closed or a television switched on. Homes are also being rapidly connected to a variety of networks, to receive visual information, to allow telephone usage or for interactive computing. Technically it is also now possible to bank and to shop from home, with the purchase price being automatically deducted from an individual's bank balance. Indeed, such facilities are available, installed and operative in a number of countries.

However, as well as having many positive features, a home which firstly can look after itself and secondly is networked with other homes to banks, shops and local authority stations has another side to it.

While information on the network can be used by individuals for their own benefit, this means that information on them will also be available on the network for other purposes. If individuals carry out most of their transactions, most of their shopping, most of their interests on the network, then it is very easy for them to be monitored. In this way information can be gained on what food they buy, how often they buy it, how often they purchase wine, what magazines they buy, what they do in their leisure time, and so on. Some specific examples of this were considered in Chapter 1. In short, a comprehensive set of documentation about an individual can be obtained very easily. It is not clear whether George Orwell[44] envisaged this form of surveillance in his book *Nineteen-Eighty-Four*. However, the concept of 'Big Brother' watching us becomes increasingly real. It is important to realise, however, that whilst machines are obtaining more and more information on humans, humans hold less and less information about machines.

Merely collecting information on a large computer network is only one aspect. What a machine can do with this data is also important. One often-quoted form of machine intelligence is that used in chess-playing machines. Chess, like music, has been held up in the past as being something which requires a certain amount of intelligence. Indeed, chess Grand Masters and top classical musicians have been regarded in society as being very intelligent people. However, just as it has been found that machines can play music better than most, if not quite all, humans, so it has also been found that machines can play chess better than most, and perhaps all, humans.

In the first instance it is worth pointing out that for humans, whilst a certain amount of raw talent, basic instinct or inherited ability – call it what you will – is required to play both chess and music, both also require a certain amount of learning and experience. It is most often a machine with mainly a gene program, with little or no learning at all, which is compared with a human and has until now, even on this basis, been found to be highly competitive.

For chess the problem is well contained, in that there are just 64 individual small squares on a chess board, with 16 white and 16 black pieces. Each of the pieces then has only a small number of ways in which it can travel, and their starting points are fixed. The machine can, therefore, be programmed so that from any situation on the board it can try to move each one of the 16 pieces at its disposal, and test out what is likely to happen, before it actually selects which move to make next. So essentially it predicts ahead: if I do this next, then this, then this, etc. etc., the result is likely to be . . . and so on to the end of the game, with the aim of winning.

Rules of thumb can also be added into the program, and possibly even learned by the computer, along the lines of the expert system discussed earlier. For example: if the Queen is threatened, then move the Queen. Some of these rules could even become dependent on the opponent's individual style of play.

When playing against an awful, twice-a-year chess-player such as myself, a fairly standard chess-playing machine should win easily. It can simply run through all of the possible moves, at any time, very swiftly and throughly, while I am bound to make mistakes or simply overlook a game plan. A good chess computer can beat the best humans, although a human Grand Master, knowing how the computer is likely to behave, perhaps because he knows how it has been programmed, can sometimes create a situation where the computer takes a more profitable short-term view only to lose in the long term. It is an interesting exercise[45] to find situations on the chess board in which the best chess-playing computers make a 'mistake'. However, as programming improves, such situations are getting few and far between and Grand Masters now lose many more games than they win against the best computers.

Essentially, a chess-playing computer of today is based largely on a gene program with little or often no learning at all. A human chess-player, meanwhile, is based on a certain amount of gene program, but a lot of experiences learned from games in the past. A human tends to learn particular moves or tricks or to recognise certain situations and to operate nice little plays at particular times. The chess-playing computer in its present form does not really have a 'feel' for the game but merely carries out preset instructions. However, even in this situation it is possible for machines to beat humans at a challenge that used to be, and indeed still is, regarded as a game which requires intelligence.

In February 1996 an IBM computer system called Deep Blue defeated the erstwhile chess champion amongst humans, Gary Kasparov in the first game of a six-match series. However, Deep Blue only managed to draw two of the remaining five games, and hence lost the series overall. Deep Blue took its defeat in good part.

May 1997 saw a 1 million dollar rematch, described as being vital for the human race that Kasparov be victorious. However, on this occasion, Deep Blue, which can perform more calculations in the blink of an eye than most humans will complete in a week, came out on top, winning 2 games to 1 with 1 drawn. At the end of the event Kasparov stormed off the stage claiming that if Deep Blue had not been programmed specifically to defeat his game plan, it would never have won.

Despite the fact that Deep Blue's victory was over the World Chess Champion, the 'best' human at the game, it is difficult to describe it as a particularly intelligent machine. It, in common with other chess-playing computers, simply looks at different potential moves, assesses the outcomes and decides on the best strategy. Its victory can therefore be seen as a small step for a computer but a giant leap for machinekind.

It could be said, in defence of Kasparov, that chess is merely a game which is well suited to computers, and that Deep Blue's victory was therefore a shallow one. Realistically however the links between chess and military strategy, particularly its psychological aspects, have been long recognised. The effect of a Deep Blue win in a battlefield scenario takes on a much greater significance.

It could also be said that Kasparov has a deeper understanding of the game as he 'knows' he is playing chess and that it is merely a *game*, whereas Deep Blue does not. In my opinion this philosophy must be overcome, as it is, in effect, putting humans on a pedestal from which we must fall, when the realisation comes that we are not the only intelligent entities in this world. The important point here is that Deep Blue won, how it did so being of minor importance, the fact being that it did *win*. As far as we know it may well have 'known' what it was doing, but from a machine perspective, rather than a human one, and just because this is not necessarily the same thing should not lessen its victory.

It would be possible to realise chess-playing computers based on neural network models, in which a much more human-like response is obtained, as opposed to the hard logic of standard computers. In these, only some very fundamental basic program elements are included, and most of the machine's responses would be learned ones, based on the experience of playing a number of games. This would probably prove to be a much more difficult route for a machine to take, however, and it may simply be 'best' to retain much of the basic, programmed style of play and to merely use a neural network for fine-tuning and 'special case' plays.

A comparison between chess-playing machines and humans does also raise the question of how chess-playing humans compare with each other. Are better human chess-players generally more intelligent than other humans, do they merely have a much better memory for the specific problems of the chess board, or have they simply had a lot more relevant experience? If we do regard a human who can play chess well to be intelligent largely because of his chess-playing, and yet he is soundly beaten by a programmed chess-playing machine, how intelligent is the machine therefore?

It could be said that in chess, as with other things, a human can occasionally respond in a completely 'random' way, which means essentially that their action is not dependent in any way on what has gone before. Despite what we might sometimes like to feel, it does not take much thinking to realise that humans never respond in a random way, in anything we do. Everything is based on our initial gene program and experiences in life. Some actions may appear random to other people, but that is simply because they do not fully comprehend an individual's gene program nor have they witnessed all of the same experiences in life.

Even if, as humans, we try to act in a random way, we use some control over what we do. Assume that I roll a die and, without anyone else seeing, I shout out the number resulting. This can be regarded, by those listening, as a random act on my part, particularly if they know nothing about dice. But realistically it is entirely dependent on how I hold and throw the die, and how it rolls. Strictly speaking, we should be able to calculate the number shown at each throw, except for the effects of chance, such as a sudden gust of wind, although even that could be modelled.

With machines, meanwhile, we can have pseudo-random

events. Pseudo because usually a sequence of events is started off which are linked to each other in a relatively complex way, but which appear to be scattered and unpredictable unless the links are known. This is, in fact, very similar to the dice-throwing exercise. Realistically, however, from the outside it would appear to make no difference at all whether an event is random or pseudo-random. In fact, if randomness in a machine is dependent on its experiences and genes, rather than on a sequence started by humans, it becomes random in exactly the same way as a human's random action, which essentially is not really random at all.

To look at this another way, assume that I select any ten numbers between 1 and 10, where some of the numbers can be the same. So I select 3 7 6 10 3 4 1 9 4 7. To someone else, you, the reader, the numbers appear to be completely random, in that each number, as it occurs, does not appear to be based on anything that has gone before; each one cannot be predicted. However, I have actually used a method to choose the numbers, but I'm not going to tell you what it is. The numbers are therefore not really random, but pseudo-random. To you it makes no difference; random or pseudo-random, it's all the same really. Of course, it could easily be a machine that selects the numbers and not me, or it could be a series of events or actions that are selected rather than numbers.

An amount of pseudo-randomness can, however, be used to aid learning, either by a human or a machine. As an example, assume that you are forced into a room from which lead five doors, one of which, the white one, is the door through which you entered and will later leave. The other doors are painted red, yellow, blue and green respectively. You are told that there is some money behind one door and you must find it, but you may only look behind one door. You are also told that the money will be behind any one of the doors, except the white one. So what do you do? Do you act completely randomly, do you act pseudo-randomly or do you use some ploy? Maybe you choose your favourite colour, or guess, using logic of some kind. What is certain is that whether you are a human or a machine, there is no advantage, the problem is just the same.

You guess wrong on the first attempt and are told that the

money has now been moved, but it is less likely to be behind the door you have already opened. If the exercise is then repeated again and again you may find that four times out of ten the money is behind the green door, three times out of ten behind the yellow door, two times out of ten behind the blue door and one time out of ten behind the red door. Now you have to solve the problem again. Which door do you look behind first? Hopefully the green one! So whether human or machine, by trying some initial fairly random searches you have gained a good idea of where the money is most likely to be. Now each time you tackle the problem you perhaps spend more time looking behind the green door, a little less behind the yellow door, and so on. You have learned that it is worthwhile trying certain tactics more often.

If, however, you start the problem again from the very beginning and this time begin with the knowledge (maybe you have read it somewhere) that in problems like this the organisers rarely put the money behind the blue door, your initial search would, I assume, not be completely random, nor would it have been as successful as if you had read that it was the red door which is rarely used. Again, the same would be true for both human and machine. At the end of it, though, you probably have a nice strategy to use.

Whilst the problem remains as it is you can apply the same strategy with reasonable success, until one day the organiser moves things round and, for example, changes his preference for the yellow and blue doors and also includes a purple door, or hides two portions of money at a time. In each case the problem has not completely changed, but it has changed enough to warrant a different ploy or to cause you to adapt to the new set of circumstances. So a further element of pseudo-random behaviour is required until the new situation has been learned.

The example given is a simple version of many far more complex problems. Strategies for buying stocks and shares, strategies for routeing electrical power round the country and strategies for minimising travelling distances all can be tackled by essentially the same approach. This requires an element of either pseudo-random behaviour or weighted random behaviour, when certain actions are more likely than others. Instead of four or five doors, however, these problems may require 100 or 1,000 possibilities, and instead of the problem changing once a day it might change every minute. Machines or humans can both tackle the problem, both can learn, both can adapt. But nowadays machines can usually adapt more quickly and can deal with a much greater number of possibilities.

This chapter has been concerned not so much with robot machines, but rather with large machines, more than likely computers, possibly connected together to form a very large network. As part of this network the main intelligence base could be at one point, possibly in a central computer, but also it could be distributed around the network: intelligence directly where it is needed, on tap, so to speak. Such an arrangement has many positive aspects to it. Indeed, it can be considered that the machines are helping humans, carrying out boring, mundane calculations more accurately and more efficiently than humans can. Storing information about humans on the network is also very useful, as it is easy to retrieve again and can be very quickly displayed in the form of colourful block diagrams, numerical values or interactive charts. Very detailed and lengthy pieces of vital information can be sent, in a split second, to the other side of the world, with no mistakes.

This is all extremely positive, and it is no wonder, therefore, that there is considerable pressure for us all to conform to this new, technological way of doing things. Each year new and more sophisticated machinery is made available, with more functions, more options and more intelligence. Not to use it means dropping behind, getting out of date, and possibly going out of business.

On the other hand, by going in this direction, humans are handing over power, control, information and decisions to machinery. Machines are filling the role of middle people and blue-collar workers. Humans in many situations are rapidly becoming dependent on machines. On arrival at the office, the first job is to switch on the computer, let it know you are there, see who sent you any messages and check what you have to do for the day.

It is clear, in the late 1990s, that humans are using machines. We are in control, we switch them on and off. But the pressure is very much in another direction, to hand over information to machines and to let machines make some of the decisions. You, the human, no longer need to do this or that, the machine will do it much better and much faster. Machines can communicate with each other, all over the world, very efficiently and very effectively. It is humans who have given them that ability and it is humans who at present use machines in order to communicate with each other, all over the world. Without the help of machines, we cannot do it anywhere near as efficiently, anywhere near as effectively.

In the way we live, humans each year become more and more dependent on machines. As more sophisticated machines are developed, so we make use of them within society, and society changes a little bit more. But what happens when a machine fails, when a central computer in a company breaks down or our automobile stops working? Often we are lost. We simply cannot operate, cannot do our job, cannot travel around, until things have been put back to normal.

Humans having such a strong dependence on machines potentially puts machines in a strong position themselves. But whilst humans are in control of the machines and machines have, at best, a very limited level of intelligence, there is not really much of a problem. In the same way that human (robot) workers could be exploited in the past, so now we can exploit machine robots, machine workers.

The key is obviously the intelligence of our machines. While computer workstations simply do as we tell them, then there is no problem. While machines communicate with each other across the world only when we tell them, then that is satisfactory. While stocks and shares are purchased only when we say so, then things are fine. But that is not what is happening *even now*. Machines are gaining the power to question what we tell them, or advise us on a better way. Machines have the power to communicate across the world when they decide it is the right time. Machines have the power to decide which stocks and shares to buy and sell and to make the transactions when they decide.

While machines are less intelligent than humans, we are still in control. How long will this situation last, though?

Chapter Seven

Human Brains Versus Machine Brains

Human brains are complex biological organisms. Simulations or attempted copies of the actions of a brain are usually made with electronic circuitry, possibly using a computer, although optical and even mechanical systems can be used. These are very different starting points and hence one can expect that, no matter how good the simulations, human and artificial, machine brains will behave differently in some ways.

When trying to model how a human brain works, so that we can copy at least some aspects of its performance, we can stimulate it and see what happens. If then we can build a model which, when we stimulate it, behaves in the same way as the original, that model will be a good simulation of a brain.

Unfortunately, human brains are very nonlinear, and this makes them difficult to model. Although things on the Earth are generally nonlinear, in some ways they are often fairly linear. Consider a bank account. If I have £5 in the account I shall be paid interest on this, whereas if I have twice as much in the account I shall be paid twice the interest. Ten times the amount gives me ten times the interest. This is linear behaviour. But if I owe the bank £5 I shall have to pay them a lot more than they would have paid me in interest on my £5. This is nonlinear behaviour.

A simple experiment can be conducted to show the nonlinear behaviour of a human brain. Buy your boyfriend, girlfriend, husband, wife or partner a present and you will in return get from him or her, let us say, two kisses. Next week give two presents and see if you get four kisses. I suspect not. Just to be sure, buy three presents in the following week and see if you get six kisses. If anyone does get this far and has obtained two, four and six kisses respectively, then their partner will have behaved in a linear way, and should be treated with considerable suspicion. Virtually *any* other response can be regarded as a 'normal' nonlinear human brain response.

To prove the nonlinear behaviour conclusively, the number of presents purchased can be built up each week until about week 5 or 6, by which time your partner will start feeling that you are cheating on him or her. After about week 10 he or she will probably be convinced of your two-timing and will leave you. In this way the experiment will have been quite costly, but you will have proved conclusively the nonlinear behaviour of at least one human brain.

It is this type of behaviour, however, which is so difficult to model artificially. The action taken in week 10 could easily be the opposite of the action taken in week 2 or 3, and yet the stimulus could be very much the same type of thing. Only the size of the stimulus (the number of presents) is different.

Just as with one human, a particular, albeit nonlinear, response is obtained, so with another human, the same stimulus would almost certainly produce a different response. A clever partner may not care about whether or not you are being unfaithful, but will have worked out that, if you continue your experiment, in six years' time he or she will be receiving over 300 presents a week. But then, would *you* behave in a linear way? Indeed, could you afford to?

The fact that humans behave and perform differently from each other makes it more difficult to compare machine behaviour with human behaviour. If we all behaved in the same way or selected a *typical* human, we would have a standard brain package against which we could directly compare the performance of artificial brains and aspects of machine intelligence. Unfortunately, a person used in this way would no longer be a typical human.

Just as humans exhibit different characteristics from each other in the way their brains respond, so too do machines, in comparison with both each other and humans. There are some ways, though, in which we are aware that machine brains can already outperform human brains. For example, machine brains can be much faster in the way they operate or make decisions, they can be much more accurate, more reliable, keep performing at the same level for a long time, deal with numerous things at (roughly) the same time, consider many different possibilities very quickly, remember facts accurately, perform complex mathematics rapidly, learn much more quickly, and so on.

But despite all these advantages for machines, at the moment humans are still in the driving seat; we have overall control. Besides, there are even now some things which humans can do that machines cannot. In particular, each human is capable of doing a wide variety of things. In saying this, we are not so much interested in physical capabilities as in the numerous different things we can do through our intelligence. Conversely, by taking an individual human capability we can usually obtain a machine not only to carry out the same task, but to do it much better.

I was involved recently in a project with an international toiletries company. The particular problem was the production of pots of a well-known cream. In the production process, once the cream was in the pot, a firm seal was placed on top to ensure that no cream escaped. It was one person's job to check on the process to make sure that the machinery put the seals on correctly. Occasionally it did not, and the person had to pull the faulty pot and seal out of the way quickly before the machinery clogged up.

In order to produce profitably, from three to a maximum of six pots moved every second along the production line past the person checking. This operated continuously. For a human it is a difficult job continuously watching pots moving past you, hoping to catch the odd faulty one.

But machinery based on neural networks, which will be described later, could be taught what a perfect pot or seal looked like and could be placed on the production line to check for a faulty arrangement. It would not get bored or tired, and even more significant than this, could deal with 50 pots a second, if required. This is just one typical example of machinery taking over a task carried out by a human because it can do it better.

In this case the machinery carries out a very rapid and accurate visual inspection check.

Most machinery carries out just one task, simply because that is all it is required to do. Other machinery has one main role although it does quite a number of things in carrying out that role. For example, an automobile acts as transport but can also keep you warm in winter, cool in summer and can even play you music. In terms of intelligent acts, though, either machines are designed to operate in one way on one thing or, in the case of a computer, a basic functionality is provided which you, the human, can arrange and program to get it to do what you want. So machines can, and sometimes do, carry out a number of tasks, although these are usually aimed at a specific overall goal.

Humans, for the most part, can and do act intelligently over a much wider range of things. However, in the case of certain humans the distinction is not so clear. Consider, if you will, the role of the actor Dustin Hoffman in the film *Rain Man*. In this film Dustin plays the part of a person with autism who, although he appears to have no *physical* problems whatsoever, is incapable of doing many things in everyday life that are usually taken for granted. In one scene from the film, some food is left too long in the oven and it begins to smoke. Dustin's character is unable to cope with this and does not consider switching the oven off or removing the food. Consequently fire alarms sound, which makes matters worse.

His character is unable to stand a bottle up correctly in a cupboard, and does not realise that it will fall out of the cupboard if it is not placed inside reasonably well. Yet, on the other hand, in another scene a box of toothpicks is spilt on the floor. By just looking at them, in a mess on the floor, he is able to say *exactly* how many toothpicks there are. In another scene he can remember facts about aircraft disasters: when, where, how many died, why the accident happened, what company it was, etc. He is also able to carry out extremely complex mathematical calculations very quickly and very accurately, to the extent that his brother, played by Tom Cruise, says of him, 'He's a genius, he should work for NASA or something.'

In many ways the character depicted by Dustin Hoffman in

Rain Man is performing in a much more 'machinelike' way than is usual for a human. He is able to deal with certain tasks, particularly those which are numerical or involve memory, far better than other humans, but is not able to cope with other jobs at all. In the film he thinks a candy bar costs the same as an automobile, and he cannot cross the road correctly. He cannot cope with many aspects of life which most people would, ordinarily, consider to require only common sense or everyday understanding.

All this prompts the question, 'Is the character played by Dustin Hoffman in the film *Rain Man* intelligent?' My own answer is an immediate, 'Yes, of course he is!' His intelligence is, however, different from a person who does not have autism. In some ways he is better, in some ways he is not so good, and on average it is difficult to tell. In some respects, as pointed out by Tom Cruise's character, he is a genius. Yet in many ways Dustin's character does not *understand* some basic everyday concepts, for example that an automobile and a candy bar do not cost the same. This lack of understanding, to some people's way of thinking[34], means that such an autistic person is not intelligent at all, on the grounds that 'Intelligence without understanding is a misnomer.' I am afraid that I cannot go along with this line of thought.

Some other autistic people are not so mentally able as Dustin Hoffman's character. Some are unable to talk. Others are unable to communicate at all with other humans. It is horrible to imagine what could be, to my way of thinking, an extremely intelligent individual being locked in a body with which they cannot communicate with others, or can only communicate in a very trivial way. Usually, the more intelligent people are, the more they need stimulation and interaction, the more they need a challenge, problems to think about and solve. Yet what does such a person get to stretch them? Probably very little in many cases.

Autism is just one condition in which a person's intelligence does not have complete access to the outside world. Physical characteristics such as deafness, blindness or dumbness can all be restrictive, but at least if a person only has one such disability, and it is merely physical, then he or she still has the possibility of mental interaction and stimulation, which must be the most important aspect. But where mental disabilities occur, this can mean that he or she is unable to communicate or function in a non-disabled way and life can be very difficult. Indeed, it is frightening to imagine just what life must be like as a highly intelligent person being treated in some ways like a baby; being clothed, fed, generally looked after and talked to in simple terms. I am saddened by the fact that often we, who can function well physically, have not been able to communicate with the person. We have not been able to tunnel through his or her brain to his or her intelligence.

The point I am making here is that machines and humans generally exhibit different forms of intelligence. In some ways they can be the same, in some ways one is better than the other and in some ways they are simply different. Even amongst humans (or for that matter machines) different forms of intelligence exist. Different people are intelligent in different ways and one person may be considered to be more intelligent than another simply due to fashion or circumstances rather than anything else. For an evening's entertainment a singer can be regarded as brilliant, whilst a rocket scientist is boring and stupid.* When it comes to designing rockets to travel through space, however, the rocket scientist is the brilliant one and it is the singer who is relatively useless.

Often a human who is regarded as a genius is exceptional in only one field, or a very small range of fields, and sometimes finds it difficult to cope with everyday life or some particular aspects of everyday life. Composers such as Mozart, Smetana and Schumann led strange lives. Mozart was considered eccentric in almost everything he did, and both Smetana and Schumann eventually went mad. Many exceptional sports people, musicians, poets and successful authors have found life outside their work to be very difficult. Several turned to drink, drugs and/or extreme sexual overindulgence and died at a very early age.

However, we do not have to shy away from the fact that machines can be intelligent, albeit most likely in different ways to

---

* Personally the only rocket scientist I have ever met was actually very good company. The example is more to make a point.

humans. It is thus not fitting to try to find particular characteristics of certain humans and say that, because a machine does not also exhibit these particular characteristics, it cannot be intelligent. If we were to do so we would also exclude many humans from being intelligent, by our definition, even though we know that in reality they *are* intelligent. Very simply, machines and humans can be, and are, intelligent in different ways, and it is only in their performance on specific tasks that we can really make a direct comparison. Indeed, this is exactly the case with humans, who can be given a particular test and ranked on the results, which relate to aspects of intelligence indicated by that test and that test only.

In creating artificially intelligent machines we often apply, in a way, a reverse test whereby, for a specific problem, the machine gives a similar answer to a human. As described earlier, for an expert system we simply look at certain circumstances and see what a human would do under those conditions. The artificially intelligent machine is then set up so that it behaves in the same way, although it will probably be more accurate and reliable.

An envelope-sorting machine was described earlier. Another example would be a home heater. Under human control, the heater could work as follows:

If the room is cold and the heater is off, then switch on the heater.
If the room is hot and the heater is on, then switch off the heater.

So a human is controlling the temperature in a room, simply by means of these two rules. There could, of course, be a whole range of other rules on which the human works, for example:

If the room is cold and the heater has been on for some time, then check that the heater is working, all the windows are closed, you have some clothes on or an alien spaceship is not sucking heat from your room.

In constructing an artificially intelligent system which controls the temperature in the room, instead of the human doing so, it is fairly straightforward to replicate the first two of the rules given for switching the heater on and off. Typically some temperature measurement can be made automatically so that, when the room temperature is below a prescribed value, the room is considered cold. Hence the first rule applies and the heater is switched on by the machine. In the same way, when the measured temperature goes above another, higher value, the room is deemed hot and the heater is switched off again through the second rule.

This form of artificially intelligent system is fairly easy to put together and operate, either by means of a computer system or realistically, as is done in most practical situations, with a couple of pieces of metal acting as a switch called a thermostat, which is an extremely simple computer. For fairly straightforward and well-defined tasks such as this, an expert system is a good way of going about it.

One of the first expert systems used was MYCIN, which was a medical question-and-answer system. Instead of going to a doctor you simply communicated with MYCIN, which would ask you questions about any physical complaint that you happened to have, for example:

Machine:	Do you have red spots?	Patient:	No.
Machine:	Do you have a cough?	Patient:	Yes.
Machine:	Do you have a high temperature?	Patient:	Yes.

At the end, the machine diagnosed your problem. MYCIN was, of course, more complicated than this, but the example given shows the principle.

Expert systems can be extremely powerful devices. However, a number of problems exist. Firstly, human experts tend to disagree over certain things, and in building up an artificial expert, taking some average value between two extreme views may not be possible. Assume, for example, that we are putting together an artificial expert vehicle driver. For the circumstances where another vehicle emerges unexpectedly from a side turning ahead, one expert says, 'apply the brakes' whereas another expert advises, 'accelerate hard'. Taking an average might mean doing nothing at all, which could have disastrous consequences.

A second problem is actually getting the relevant information from experts. Either the experts do something fairly routinely and simply never tell anyone, or the language they use in describing what they do is so full of jargon that it is difficult to understand what they are talking about.

Other, perhaps more serious problems occur with expert systems. One is that rules are required in order to cover *all* eventualities, no matter how unlikely. This can mean that millions of rules are required for a fairly trivial problem, and a computer may take considerable time to check each rule for every problem. In comparison with MYCIN, a doctor must cope with almost anything when he is being consulted, whereas it would be difficult to achieve such a diversity solely with an expert system.

One further problem is that, by and large, we only need an artificial expert to investigate common-sense, sensible things, and this requires some basic understanding of what is and is not sensible, which is difficult to obtain. As an example, for the room temperature controller described earlier, when the room remains cold, it *is* common sense to check that no windows are open, but it is *not* common sense to check for alien spaceships every time.

What is and what is not common sense is something that, as humans, we learn over our lifetime, and hence we are able to put things in context or make quick decisions based on what we feel is normal or usual. The entire gamut of common-sense things is necessarily extremely difficult, maybe impossible, to program into a machine, other than for a very tightly defined problem, and even then it is not likely to be complete. The only way that an expert system would be able to make its own assumptions and decisions on a broad array of common-sense things would be if it is allowed to learn for itself, then to try things out and thereby gain its own basis of common sense.

While it is quite possible for an expert system to learn, as will be described shortly, it is not yet realistically possible for a machine to have all the same types of experiences that a human would have, in the way that a human would, as the machine would essentially have to *be* a human to do so. So, in reality, a machine's common sense would be based on the area of operation on which it is working and learning. So an expert system controlling the temperature of a room is very likely to experience doors and windows opening, and can learn from this, but it is very unlikely, while doing so, to learn about a helicopter which is cooling the room by hovering just outside.

In the same way, if we, as humans, were just controlling the temperature of a room all our lives, and never did anything else, we would not know what a helicopter was and what its effect could be. Unless we learn about common-sense things ourselves, we do not know how to deal with them. Nevertheless, humans and machines alike can make a best guess, and even if we have not encountered a particular event before, we can try to deal with it by extrapolating from what we already know. So if, as a human, we are approached in the street by a 'Muglug' (author's invention), our reaction will entirely depend on what it does and how it reacts to us. If it is small and black and furry and makes sweet noises then we might stroke it, whereas if it is 12 ft high and is brandishing a machine-gun then we probably would not.

A common-sense, reasoning approach is sometimes twisted to work against us. Many magicians' tricks are simply cases of the audience expecting to see or hear something in a particular situation and being fooled because something entirely different is actually happening. Many tricks of perception exist. For example, straight lines can appear to us to be curved because of the shape of the lines surrounding them.

Learning occurs in basically two ways. On the one hand the thing, maybe human, that is learning can simply be shown or told something or may witness something. This is passive learning. On the other hand the thing or person can actually do something or cause something else to happen and learn from observing the result, so the thing or person has itself initiated the learning. This is active learning.

Assuming for a moment that a person is learning, then different people learn better in different ways. Some are more content with being told facts which they memorise or statements which they readily accept. Others, meanwhile, learn better by trying out a possibility and seeing what happens. Some people do not easily believe what they are told and must inevitably try it out if they are to be certain. For almost all people, though, it is a combination of the two, passive and active learning.

Many factors affect how well we learn, or how readily we accept information. Examples of this are *who* teaches us, in that we accept the word of one person but disbelieve another, *when* we are taught something, *how* we are taught it, what we are feeling like at the time, whether we are thinking about something else, and so on. What our level of understanding and knowledge is at the time of learning is critical. If we are unable to put what we should be learning into context, then it is difficult to comprehend. For example, if told that when travelling from Boston to New York it is quickest to go via Pittsburgh, someone with no geographical knowledge of the area might simply accept the statement as fact, whereas someone who has relevant knowledge would most likely say, 'What a load of rubbish.'

Just like humans, machines can also learn in two different ways. In passive learning, the effect is to make an addition to a machine's program, through either extra rules or related data. The same is really true of humans. When we are given many facts and expected to memorise them, we are in effect being programmed further, although this must necessarily be seen as an upgrade or modification of our basic gene program. Exactly the same is true for a machine. It must be remembered, of course, that such a machine must have the ability to learn in this way. Many do, but this is not always the case. However, humans almost all have the ability to learn in this way.

It is also quite possible for a machine to engage in the active, experimental type of learning, although it does need something with which it can poke, probe or question in order to find out things. Either a question can be posed and an answer obtained, or something is stimulated and a result witnessed. A human can learn about gravity by cutting an apple from a tree and watching it fall. To attempt to learn the same thing, a machine must firstly have some basic knowledge of trees, apples and gravitational forces, but it is also useful to have a purpose behind the experiment and the ability to learn a sensible conclusion from the result.

Let us assume, for some purpose, that a machine is trying out the following:

1. If the apple falls downwards, then gravity acts towards the Earth
2. If the apple remains where it is, then gravity does not exist
3. If the apple rises upwards, then gravity acts away from the Earth

Assume also that when the machine carries out the experiment, rule 1 is found to apply. Does this mean immediately that rule 1 is correct and rules 2 and 3 are not? Maybe so for this particular example, but it might take several experiments before a conclusion can be drawn. A better example might be: 'If I boil an egg for four minutes, the yoke is still runny.' Perhaps when you conduct the experiment, seven times out of ten the yoke is runny and three times it is not. Presumably you then either boil each egg for 3½ minutes (if you want a runny egg!) or risk the fact that about three out of every ten eggs you boil will not have a runny yoke.

Now, assume that you carry out the apple experiment and find that, with no tricks, once the apple has been cut from the tree it remains where it is, i.e. rule 2 applies. What do you think about this? Is it magic? In this case, common sense dictates that in normal life rule 1 *must* apply. If not, there must be some exceptional circumstance. This is a problem for humans with common-sense reasoning, as we can erroneously end up believing in something with 100 per cent certainty. There are many things that could keep the apple where it is, but because they do not *normally* occur, we can believe that they will *never* occur.

One interesting discussion around the topic of machine intelligence, the frame problem[38] (what is relevant and what is not in a particular frame, and how broad that frame is), is to my own mind enormously misleading. See what you think; it goes like this:

A robot has been built with the sole task of fending for itself. It finds out that its precious spare battery, the only one, is locked in a room in which there is also a time bomb set to go off soon. The robot finds the room, opens the door and pulls the battery out of the room, by means of a wagon, before the bomb goes off. Unfortunately the bomb was also on the wagon. The robot knew that the battery was on the wagon but did not realise that pulling the wagon from the room would also pull out the bomb.

So the designers build another robot and it also tries to tackle the problem. This time the robot considers the potential implications before it pulls the wagon out of the room. It has worked out that moving the wagon will not change the colour of the walls and is starting on a set of irrelevant calculations when the bomb goes off. And so it can go on, with further designs of the robot and further programs.

In real, practical expert systems, such a situation is indeed an enormous problem. In respect of this problem, let us give the robot a full, succinct understanding of bombs and relevant consequences. Now the robot deals with the bomb, but falls through the trap door we have inserted. So we give the robot an understanding of trap doors and it avoids the trap door but walks into the giant magnet we have just lowered from the ceiling. The point is that if the robot has not had a lifetime's experiences of these things in a human world, then it will never perform as well as a human, and we will never be able to program it to deal with *all* of the common-sense things it needs to know. We would, I guess, have the same problem with a human who had not experienced life.

I certainly agree that the frame problem is a practical, real-world problem for expert systems and related artificial intelligence methods. However, I do *not* agree that this is a decisively limiting factor for machine intelligence. It is biased, like many arguments from a human's point of view, to make humans look better!

Assume that *you* have to go into the room with the battery and bomb and pull out the battery. In the room is also a Muglug, who says, in perfect English, 'If you move that battery I will destroy you.' So what do you do? What would the robot do in this situation? Now you are in the same situation as the robot. You are both suffering from the frame problem. You both need common-sense knowledge about the Muglug before making a decision. I now tell you that the Muglug is just an academic philosopher, and he will not really kill you: he is just saying that to see what you do. So you take the bomb off the wagon, prior to rushing for the door. Unfortunately the bomb also has a trembler detonator on board, and when you move it, it explodes. Sorry, I did not tell you that!

The frame problem itself, although a major problem for machine intelligence of the expert system type, is not a problem specific to machines. It is a problem for humans and in fact all other creatures as well. How do we deal with a situation which depends on what relevant information we have and what common-sense reasoning we can apply to it, as well as basic instincts which may help?

Expert systems are, in their fundamental form, a logical approach to artificial intelligence, obtained by looking at the brain from the outside, seeing how it behaves and attempting to copy this behaviour. There are many other approaches. Expert systems do relate closely to how humans do things. But do we do everything in terms of 'If . . . then . . . '?

It is fairly easy to learn using expert system rules, although it does present a few problems. Firstly, if the system learns, then it is no longer copying an expert in whatever it is doing, but rather it is doing what it has learned to do. Also, the system can keep trying different rules in a situation. If they produce success then it is more likely to use them next time, whereas if they fail it is less likely to use them next time. However, some rules need to be used only very rarely, although they are vitally important at certain times. So the question is, how many rules does the system retain? If it keeps retaining all rules, it could take quite a long time to see which rule should operate at a certain time, simply because so many rules have to be checked.

Usually a fixed number, a population, of rules is retained, and when a new rule proves to be successful, it is kept and a less successful rule is removed, whilst attempting to retain vitally important rules. Assume that we have an expert system to drive an automobile. Presumably we would like a rule that stops the automobile when a person runs into the road in front of it, even though this rule would only be needed on rare occasions. Meanwhile we could use a lot of rules to deal with ordinary driving: cornering, accelerating, slowing down, and so on. There is, therefore, a trade-off between the number of rules retained and the speed of operation, and another trade-off between the number of well-used rules retained and the rarely used, but important, rules.

Some of these trade-offs also have to be made with humans. Consider the example of driving an automobile. Because we have to take into account the possibility of someone stepping into the road in front of us, we must drive carefully and slowly. If we did not bother with such a possibility, we could drive much more quickly. The everyday performance would improve if we did not need to bother with exceptional cases.

An expert system can learn, either by being directly taught or reprogrammed, or by trial and error. The first approach, passive learning, is obvious. The second, active learning, is less so. Each rule held can have a strength or score associated with it. When a situation occurs and a rule is tried, and found to be successful, its score can be increased. If it is not successful, its score is decreased. What is required is some way of measuring the success of a rule, and an automatic critic to increase or decrease its score. New rules can be generated, perhaps pseudo-randomly or perhaps by mixing up other rules, using genetic methods, and subsequently tried out. If they do not prove to be successful, their score will drop and they can be removed. If, however, they are successful, they can be retained. If we look at our expert system for driving an automobile, we could, for example, have a set of rules to deal with a left-hand bend:

1. If there is a left-hand bend, then slow down rapidly, steer sharply to the left, change gear.
2. If there is a left-hand bend, then slow down slowly, steer gently to the left, change gear.

The system could be taught by a driving instructor critic, who for a particular corner marks positively or negatively whichever rule was applied, so that next time a left-hand bend is approached, one rule or the other is more likely to be used.

With the automobile driving example, a big problem with expert systems can be clearly seen. We have many different types of left-hand bends, we have many situations with other traffic, roundabouts of different sizes, traffic lights, and so on. There are many situations which must be accounted for on a normal road. The number of rules required to deal with them is enormous, and even then the system must contain, and be able to use immediately, certain special, one-off rules.

We can say that an expert system might make a poor driver, but it must be remembered that this would be a machine-based system operating in a human-based environment, to which it is not well suited. If it was decided that transport would be replanned to operate in a way that is better for machines, then the situation would probably be very different. Presumably there could be rapid communication between vehicles, communication between the road and vehicles, and much more processing within a vehicle. Indeed, many systems do exist now in which an otherwise autonomous vehicle tracks a buried cable or magnetic line; furthermore, systems are available whereby one vehicle can sense a vehicle in front of it, on the road, and does not then get closer than a preset distance.

Expert systems and other artificially intelligent methods based on modelling a brain's behaviour from the outside can achieve a certain amount of success in well-defined fields, but it is certainly difficult to conceive of such a system learning, from an initial gene program set of rules, a complete mode of operation for a problem. Expert systems are essentially a good method with which to simply state some operations of an intelligent system, whether it be in terms of a human or artificial brain. What is really needed is something more, a different approach, and this is provided by neural networks.

With neural networks an attempt is made to understand, roughly, how a brain works from the inside rather than the outside; or perhaps one should say at a much lower level. It is estimated that a typical human brain has over 100 billion neurons, individual cells. Each neuron is connected with other neurons, some being connected to only a few other neurons, others to hundreds and some to tens of thousands. The way in which the neurons are connected together is also very complex and can even appear to be random, but is, of course, part of the gene program make-up of each individual.

At this stage, our knowledge of how the brain's neurons work is limited. It is very difficult to carry out investigative work on the brain's micro-elements, due to the strong possibility of damaging or even killing the elements. However, we do know something! We know that neurons in the brain are not one standard size and shape, and are not connected together in a nice neat way. We know also that certain groups of neurons, of a similar type to each other, exist in specific areas of the brain and deal with specific jobs. One group of neurons deals with our visual system, another group controls our legs, arms, etc. (motor neurones), and a further group deals with mental processing. As mentioned earlier, certain groups of neurons are concerned with short-term memory (a close link being formed here, as described earlier, with those involved in epileptic fits), while others deal with the retinas of our eyes.

All neurons tend to operate in the same sort of way. There may be bigger neurons, shorter neurons, longer ones and smaller ones, but they have, roughly speaking, a common mode of operation. In some ways neurons are just like other body cells in that they have a central nucleus and a membrane.

By looking at one neuron in isolation, we can get some idea of its operation. The overall working of the brain is, though, due to 100 billion of them, interconnected. A typical neuron has on one side of it some short, spiky, hair-like bits called dendrites, whilst on the other side it has a longer thread called an axon. Messages are sent to a neuron through its dendrites, and, dependent on what message it receives, it then fires an electrical message out through its axon. The axon of a neuron is connected to a dendrite of another neuron by a synapse, which contains a gap across which the electrical message travels when the axon has been excited. The message actually passes across the synapse by means of neurotransmitters, which are, in reality, small amounts of chemical fluid. Due to the electrical messages it receives, each dendrite has a level of electricity, a voltage. The more messages it receives, the higher its voltage.

The basic operation of a neuron is as follows. At some point in time, the dendrites of a neuron each have a voltage on them. All of the voltages going to one neuron through its dendrites then add together and the total affects the rate at which the neuron sends out signals through its axon. Some of the messages are positive and some are negative. Each neuron, however, itself only sends out either positive or negative signals but not both (though there are some exceptions to this).

When we use our eyes, so the light and dark regions and the different colours we see are converted into voltages on neurons connected to the retina, at the back of our eyes. These neurons then feed their messages into the brain, and so what we see enters the brain via the retina neurons. As what we are looking at changes, so the voltages on the dendrites change and hence the rate at which the retina neurons send out their messages will change.

As our brain operates, in whatever it is doing, so this involves neurons increasing or decreasing the rate at which they are sending out messages. Particular activities result in certain areas of the brain sending more messages at a faster rate, or with much wilder swings in the rate at which messages are being sent.

With epilepsy, when a fit occurs, normal brain operation is replaced by synchronised and repetitive electrical discharges, i.e. the voltages build up and then suddenly drop, and do so repeatedly. It is rather like, instead of having a normal telephone conversation, hearing only a very noisy crackling, so that normal conversation must stop. The epileptic situation, which is temporary, can be started off by a number of things, such as strobed disco lights or a flickering television screen.

An example of this type of thing occurred in Japan in December 1997, where more than 700 youngsters were treated in hospital for epileptic-like seizures after watching a popular cartoon series based on Nintendo's Pocket Monster games. The programme included a depiction of an explosion followed by 5 seconds of blinking red eyes flashed by the show's lead character. The particular frequency of flashing was enough to cause the outbreak.

Due to the number of children involved, the sequence was also shown, the following day, as an item on the main television news. Not surprisingly, as a result of this, further children were admitted to hospital. For some time, video recordings of this cartoon caused further admissions, with all such children displaying the same symptoms. In the UK however, legislation is in place restricting a range of frequencies of repetitive signals from appearing on our screens because of this potential problem. This means that there are a number of programmes from both the United States and Japan, for example, which can never be shown on UK television, as they contravene this regulation.

Memories and learned responses are, it is believed, stored in the brain, in the synapses, the connections between the neurons. This then affects the strength with which messages are passed around the brain. As an example, let us assume that we have learnt that when driving, every time we see a green traffic light we should press our foot on the accelerator. This association is recorded in our brain synapses, so that we respond in the appropriate way. So, there are some things we learn to do virtually automatically and which become almost a basic instinct. We are conditioned to respond directly in a certain way.

Perhaps the most famous example of a conditioned response is that arising from the work of the Russian physiologist Pavlov. In his experiments with dogs, the normal stimulus causing saliva in the dog's mouth, food, was linked with a bell ringing. So every time the bell rang, food was provided. The dogs, therefore, associated the ringing bell with food and began to salivate when the bell rang, rather than when the food appeared.

With humans, many forms of advertising actually make use of associations and conditioned responses in order to sell a product. We are conditioned to want natural-looking, glossy hair. This can be obtained if we wash our hair with product X. In particular, film star Y uses product X, and just look at her hair. So we associate product X with nice hair and are more likely to buy the product.

It is the synapses in our brain that make these associations, and cause us to respond in the way we do. Some of them are set up at birth, due to our gene program, and endow us with our basic instincts as well as, perhaps, giving us certain habits and maybe even certain memories. Alternatively the synapses operate

155

in response to what we learn, how we are conditioned and what we experience.

But how do we go about modelling such things as neurons? Indeed, are they at the *right* level of complexity? With expert systems and classical artificial intelligence, the operation of a brain is modelled and copied from the outside. By modelling at this level we can mimic some of its operations, but there are many things going on inside it that we simply cannot capture, and therefore our model is limited. If we go down to the neuron level, however, we are in one sense likely to be able to achieve some similarities. On the other hand, it is very difficult to conceive, without detailed plans of the brain, that we could reproduce everything in a very similar way.

Consider again, if you will, that we are trying to build a copy of a house. If we look from the outside of the house we would be able to obtain some information about it, and hence copy some of its features, but many things we cannot copy because we are not looking inside the house. If we now look at the bricks from which the house is made (analogous to the neurons in the brain) we can better understand how the house is put together, but without detailed plans it would be extremely difficult to reconstruct the complete house. However, we could perhaps build a wall or a small shed and learn from this.

We could look at the house even more deeply, perhaps at the clay molecules from which the bricks are made; or less deeply, at individual rooms or groups of bricks clumped together. There is essentially a wide range of levels which we can consider. But looking at the level of bricks seems to be sensible, in that we know that houses can be built up from these, and they are relatively easy to deal with.

In terms of the brain, there are those who feel it is best to look at what the neurons are themselves made of. Some have suggested that we should consider a theory such as quantum mechanics. There are others, however, who feel that it would be best to look at groups of neurons rather than individual ones. My own belief is that we can get a long way by looking at the neurons in the brain and modelling them. Just like the bricks in a house, they are a sensible level, at present, to model and try to copy. But the human brain is a very complex network of neurons and we do not have any overall plans available, so we should not, in the near future, expect to be able to put together anything that is very similar to a human brain.

The neurons in a human brain, and the way they are connected, have evolved over millions of years. We still have a lot to learn about these connections. However, we should be clear that that is *all* there is in a brain. All our thoughts, likes, dislikes, consciousness and self-awareness arise in our brains, which are neurons connected together with dendrites, axons and synapses.

Animals, birds, insects and other creatures have smaller brains, with fewer neurons and simpler connections. Their brains operate, therefore, in simpler ways. It appears that dogs and cats dream in some sense, and appear to be self-aware, but what abstract thoughts do they have? We do not really know. Meanwhile the life of an insect, such as a bee, is fairly predictable. It is as though it has simply its basic instincts, its gene-programmed brain and little more. The number of things a bee can do is limited. The way it behaves is limited, but that is not surprising as its brain is much, much smaller and simpler than that of a human.

Is a bee conscious? In a beelike way it probably is, just as a dog is conscious in a doglike way and a human is conscious in a humanlike way. Exactly the same is true of intelligence – it depends on the species. There are, of course, some similarities between the species, but also many differences. Yet it appears that dogs are more intelligent than bees and humans are, on the whole, more intelligent than dogs!

Let us take the operation of a single neuron and model it, attempting to copy an individual neuron in a human brain. If our artificial neuron is reasonably similar in its performance to a biological one, then by connecting quite a number of artificial neurons together we should be able to realise some brainlike characteristics, many of which are not exhibited by standard computers. A result of this is to increase the range of usefulness and capabilities of machines.

There are several voltages (numbers) going into the human neuron through the dendrites, and a rate of firing electrical pulses coming out of the neuron through the axon. We know that, because of the synapses, some of the numbers going into a neuron are more important than others. We also know that the relationship between the rate of firing and the inputs added together is not a straightforward linear one. We know that there are many different neurons in a brain and that they are connected together in an extremely complicated way.

But we can very simply construct an artificial neuron, either in a computer or as a piece of electronics, by saying that one neuron consists of several numbers or voltages input to the neuron, and that the number or voltage coming out of the neuron is related, via a simple nonlinear term, to the addition of the numbers coming in.

This nonlinear term could operate in the way that when the sum of the numbers coming into the neuron exceeds a defined level, so the neuron output will give one value, whereas when the addition is less than the defined level, it will give a different, possibly lower, value.

Rather than choosing directly from two set values to put out from the neuron, we could make it much more realistic by smoothing the relationship. As an example, assume that when the numbers put into a neuron add up to ten or more, the number put out is 6, and when the numbers put into the neuron add up to four or less, the number put out is 2. Now, in between, we could put out mid-range values proportionally. The most commonly used relationship is a sigmoid shape, which approximates to the one-value-or-another approach but, to an extent, also allows the output of intermediate values.

When a human brain learns, synapses cause the values put out by one neuron to have a greater or lesser effect on the input to the next neuron. In the same way, we can *teach* an artificial neuron by weighting, to a greater or lesser extent, the values put into it.

Let us assume that we wish to teach our artificial neuron a few numbers, so we put in three numbers 3, 1 and 2, through three dendrites to our neuron. We then adjust the weighting (synapses) separately to each of these dendrites, so that the value put out from the neuron is 8. Perhaps when the numbers put in are 1, 1 and 1 or 1, 3 and 2 the output must be 2, so we adjust the weights again in order to achieve this. And so it goes on, until particular patterns of numbers put into the neuron cause the value put out to be high, whereas others cause it to be low, and some give a value in between. If now we put in any three numbers, the neuron will show whether or not it *recognises* our numbers by its output value being high, low or in between.

If the values put in are much more complicated, then we need a lot more neurons, connected or networked together. As an example, consider a picture in a video camera, and for simplicity assume it is just in black and white, not colour. The picture can be split up into very many small squares – pixels. For a black and white picture of your face, each pixel will be either white, black or a shade of grey. If we associate the number 0 with white and 10 with black, then the levels of grey in between can be given values between 0 and 10; 1 for a pixel that is almost white, 9 for a pixel that is almost black, and so on. In this way the picture of your face is made up of a huge array of numbers representing the greyness of each pixel.

The values from a number of the pixels can be put into a neuron. Let us say four pixels, which for your picture have values 3,1, 7 and 6. The value put out from this neuron is then made to be high by adjusting the weights on each dendrite. The next four pixel values from your picture are put into another neuron, and so on, until all the pixels from your picture have been put into one neuron or another, each neuron dealing with four pixels. The dendrite weightings on each neuron are adjusted so that all of the neuron outputs for your picture are high values.

Now if we show a different picture to the neurons, say the picture of a tree, most of the neuron values put out will be low because the black, white and grey pixels which indicate a tree are different from those which indicate your face. Some will probably be high, though, due perhaps to similarities in the background.

What we can do is look at all the values output. If most of them are high then it is probably your face that the camera is looking at. If not, then it is probably a picture of something else. However, due to different lighting, the wind blowing your hair or even a different smile, it is unlikely that the values put out from *all* of the neurons will be high for future pictures of your face.

In this way a number of neurons can be connected together

very simply to recognise a picture. The picture could easily be in colour, but this makes things more complex. It could also be a picture of absolutely anything. So the network of neurons could be taught to recognise a particular face, or a packet of cornflakes or a book; whatever we like. If something exactly the same is later shown to the network, most of the values put out will be high; if something very different is shown then most of the output values will be low, whereas if something similar is shown then quite a few output values will be high. The network is, therefore, not only recognising one particular picture, but also giving an indication of similar pictures. It is, however, doing just one job: recognising a picture.

In the picture recognition example, training the neurons by adjusting the dendrite weights appeared to be not too difficult. In reality, though, it can be quite a problem, particularly when we consider neurons connected together to form several layers, with the values put out from the neurons in one layer being put into the neurons in the next layer, and so on. Then the weightings selected in one layer directly affect the weightings in the next layer. Neural network research has, therefore, concentrated on the use of relatively few neurons, in only one, two or at most three layers, with simple connections between them. In this way networks can be arranged and taught for specific problems and we can have some understanding of what is going on.

So where are we with neural networks at present? Firstly, neural networks made artificially are usually electronic or simply numbers in a computer, although other forms, such as optical, are apparent. Meanwhile the human brain is biological, with electrochemical signals. Secondly, each artificial neuron is usually a standard, simplistic model of a real neuron, and a network is formed by using identical neurons. In reality, human neurons are more complex, different in operation, different in size and carry out different functions. A typical human brain has something like 100 billion neurons. At present, a typical artificial neural network has maybe 10, 40 or perhaps at most 1,000 neurons. Finally, the artificial neurons are connected together in a nicely structured, layered way, so that we can understand what is going on. A human brain, however, looks a complete mess in terms of the connections, which are neither nicely symmetrical nor all of the same size.

Where does all of this get us? Is it that artificial neural networks are nothing like human or animal neural networks, therefore they cannot do the things that our brains can do, and therefore they will never be as intelligent as humans?

Some of this is true, but the conclusion is not. Artificial neurons are not exactly the same as human neurons. In theory it is possible for the artificial form to get very close to the human version. However, in reality there is presently a big gap. But the important point is that what we have are artificial neural network brains that are different, and perform differently, from human brains. As a result, machine intelligence is different from human intelligence.

The interesting question is, therefore, not so much whether we can get an artificial brain to operate in a similar way to a human brain, but rather, just what can we do with an artificial brain? What can it accomplish?

If, as before, we liken a human brain to a house and liken the neurons in a brain to the bricks making up a house, then if we can reasonably model a brick, we can try to build something with it and a few more like it. We do not have the plans to build a house, so let us build something else, maybe a shed but perhaps an office block or even a skyscraper. A house is not the ultimate in buildings. It is not a limit which demands that *all* subsequent buildings have to be smaller and simpler. So what is the limit? How high can buildings be? How many rooms can they have? How complicated can their structure be?

Similarly, with artificial brains we are looking not so much at trying to copy one particular type of structure, a human brain, but rather to construct something very different. In this way we can make use of the advantages offered by a machine brain in comparison with the limitations imposed by a human brain.

We can put bricks together in many different ways, but only in a few of these ways do they form a sensible structure such as a house. With artificial neurons too we face this problem, in that we must put them together in a sensible and systematic way to achieve something useful.

With a human brain we have something of a particular size, with a particular number of neurons, and an associated level of intelligence, with all that this means. In 20, 30, or 50 years' time humans will probably still have roughly the same size of brain, with roughly the same number of neurons, and roughly the same associated level of intelligence. Certainly we can use machines, especially computers, to help, but we are very much restricted in our development because of our lack of future physical and mental advancement.

With a machine brain we have a variety of sizes, with a relatively small number of neurons and an associated level of intelligence. But machine brains are not limited in the same way that human brains are. In particular, while human intelligence is fairly constant from year to year, machine intelligence is improving rapidly and there is no apparent theoretical or practical limit.

With a human brain we have our analogy with a house. With artificial brains we could achieve skyscrapers. In fact, there is nothing whatsoever to stop machines being not only *as* intelligent as humans, but *more* intelligent. Is this practical, though? Is it sensible? Could it *really* happen?

The aim at Reading has been to get small artificial brains to do a few things in the real world. Gradually, their capabilities are increased and they become more and more intelligent. The work described in Chapters 9 to 11 involves a successful program, using progressively more intelligent machines, at first set to do specific tasks and subsequently given a range of tasks with a variety of goals. As you will see, comparisons can, quite correctly, be drawn between machine brain behaviour and human or animal brain behaviour. After all, it is just a different form of intelligence.

The work described in this book, therefore, is based on an understanding gained from practical experience, as opposed to simply a theoretical or philosophical study. My conclusion from this is quite simple. There is nothing whatsoever to stop machines being far, far more intelligent than humans. In the chapters that follow we will look at why I come to this conclusion, in terms of how technology is changing and also in terms of the research programme carried out at Reading.

Chapter Eight

Humans Are Best?

There are ways in which humans are intelligent in which machines can never be. The state of human consciousness can never be achieved by a machine. Machines may be good on particular things but they will never be able to do everything that a human does. Humans have a special ingredient which sets them apart from all others, machines and other animals alike.

These are all very comforting statements, and as a human myself it would be nice if at least some of these statements were true. Indeed, some of them may be, but we have no scientific way whatsoever of proving any of them. In the end they may, in fact, all be true, but it might not do us humans any good in the long run. I suppose that what we *really* want to say is that humans are actually *better* than any other creature on the planet, whether it be animal or machine; that we are *more* intelligent. But is this true? I remember when I was young that humans who could make fantastically difficult and rapid mathematical calculations were said to be intelligent. So too were people who could memorise facts and repeat them when asked. On the BBC programme *Mastermind* the winner, the person who accurately remembers the most facts, is applauded and hailed as a very clever person. However, we know that machines are far better than humans could ever hope to be in terms of both mathematical calculations and memory, and, even worse, machines are getting better and better, whereas human improvement – if indeed there is any – is extremely slow.

So to an extent, our definitions of intelligence have shifted. At the turn of the century dictionaries often included aspects

163

such as memory and mathematical ability in their definitions of intelligence, but now they tend not to do so. Some dictionaries of about ten years ago included such things as learning, but now that too is disappearing as it becomes known that machines can learn just as well as humans, and in certain circumstances much better.

Despite this, many school examinations are surprisingly still based on repeating a memorised fact or multiplying numbers together. Students doing well in such tests are then regarded as being intelligent and are said to have a high IQ (Intelligence Quotient). If this is true our colleges and universities should be full of machines, because by these criteria human IQs are far inferior to those which can be accomplished by machines. In fact, by conditioning students to pass examinations, by getting them to respond in a set way to specific stimuli, we are really getting them to behave in a programmed, machinelike way. We are trying to make humans behave more like machines. We say that those who do so are intelligent. Isn't there something wrong with this?

I have a Teletext feature on my television at home, and apart from using it to find out what the weather will be like or to catch the latest football results, I can also try a Mensa IQ test which is changed regularly. Usually a whole series of numbers or words is shown and a set time is given for you to fill in some missing words or letters. If I can answer the question within ten minutes, so my television tells me, I will have an IQ in the top 4 per cent, which means that I am a particularly intelligent person!

So, I have a go at the test on the television, and manage to do the problem in not ten minutes, but only seven. Wow, what a clever person I must be! That night I am really pleased with myself, and am generally big-headed and obnoxious. After all, I am a very intelligent person; the television told me so.

On the other hand, perhaps I still have not succeeded in doing the problem after 20 minutes, so to ease my curiosity I check the answer. Obviously the answer given is completely stupid, or the question was badly worded or simply silly. Whatever the excuse I make up for my inability to solve the problem, that night I am generally annoyed with the people who set it, with the television, with everyone around me, and I am again obnoxious.

So what conclusions can be drawn from Mensa IQ tests? Firstly, just doing such tests makes me, for one reason or another, obnoxious. Secondly, I know very well that the problem it took me seven minutes to do, if I managed it at all, could probably be done by a machine in less than one millionth of a second. Maybe a very slow machine would take up to one thousandth of a second. So how intelligent does that make the machine? What IQ does it have?

We have to be very careful what we consider in general to be signs of intelligence. In practice we usually mean signs of human or, at a stretch, animal intelligence. However, this has given us considerable cause for concern as machines have become more and more powerful during this century.

Many human behaviours, such as displaying an excellent memory, were once regarded as intelligent acts, and yet now, in some ways, they are not so considered. Because we now know that machines can do much better, the characteristics are termed 'computational' and are therefore not intelligent, particularly if it is a machine which is operating. Is it that we feel that these characteristics show intelligence when a human displays them, but not otherwise?

Decision-making, as well as learning, is something that was once considered to be an intelligent act, but now we know a machine can do it better, so again it cannot be an intelligent act! Other human features, such as creativity and emotion, are on the edge of being regarded as things that machines can encompass, as we will be considering, although it is not clear why we should want an emotional machine. If we humans are 'creative' then, based on our original gene program and what we have learned, we may originate something, introduce something new or do something that no one else has thought of before. A machine is equally able to originate something. In fact, there have already been many such cases, simply because a machine can work through different possibilities much more quickly and accurately than a human. Examples are new musical tunes[46] or, in a more industrial setting, new ways of restoring subscribers to a utility when a supply fault occurs. Machines are also equally able to do things or produce results that no human has thought of.

Emotion in a machine is more difficult to imagine. However, we are simply looking at something that originates in a brain, sometimes as a feeling of pain or pleasure, often resulting in a physical action; for example, crying in response to a stimulus such as the death of a close relative. In many ways we are looking at a conditioned response which, for example, a human actor can achieve just by thinking about a stimulus. That is to say that a response can be triggered by an imagined stimulus. In the Reading University robots, which will be described shortly, the robots move away when something nasty, a predator, approaches them. Is this fear?

Picture yourself at home, alone, on a dark winter's night. The wind is howling outside, it is raining, the electricity has failed so you have no television and only candles for light. It is late. You hear a sharp 'knock, knock' at the front door. You approach the door, and a dark, shadowy figure, much bigger and taller than yourself, stands outside. You might well be unhappy with this situation, a little scared perhaps, maybe very frightened. As a result you may not open the door, or if you do, you are ready to run like mad if the person jumps at you. But why?

We are frightened, scared, happy or sad in certain circumstances, and either these emotions are inbuilt through our gene programs or we learn to feel that way as a result of our experiences. For example, the original instinct of being frightened in certain circumstances is easy to understand. Consider the case of ten humans who live in a jungle where there are many tigers. Five are scared of tigers and run away when one is near. The other five are not scared and do not run away. It is my guess that fewer of those who are not scared will survive to have children. So, in the next generation of jungle-living humans, probably a higher proportion will be scared of tigers, as this characteristic could have been genetically passed on. This is one aspect of genetic evolution.

Humans can comprehend emotions such as fear because we think we understand, by and large, what other humans feel, even if we cannot be sure of each individual's exact feelings.

Assume a tiger now approaches a mobile robot machine, and the machine responds by moving away. What does this mean? It is difficult to conceive that the machine feels emotion, in this case fear, in the same way that a human does. But nevertheless the machine is behaving in a similar way to a human and therefore can be considered to be showing machine emotion. The machine's feelings, an internal state or indication, may well have been programmed by ourselves, but, there again, a human's feelings have in a similar sense been programmed biologically by the person's parents. To say 'it's a machine, it can't have emotions' is simply not scientifically correct.

Historically, some humans have thought of other humans of a different race or colour as not having emotions or feelings (now we *all* think differently), or at least that some people's feelings are not as strong or 'cultured' as others. Does a dog or cat have emotions? You would probably answer 'Yes.' Does a fish have emotions? What about a bee? Maybe they do, maybe they don't – what do you think? I believe they all do, to a greater or lesser extent. The feelings or emotions of a bee are probably much simpler and more basic than those of a human, but then the brain of a bee is much simpler and more basic.

The whole brain function of a slug or a small insect can be artificially simulated fairly well by means of an appropriate artificial neural network, essentially because of the relatively low number of neurons involved. Indeed, certain human functions and responses can also be simulated reasonably well, even though the number of artificial neurons used may not be as high as the human would use. So, if an artificial neural network is basically performing the same function in almost the same way as the original biological neural network, then similar signals can be discovered in both, albeit in one case they may be electronic signals and in the other electrochemical.

I feel that there is no way, at present, in which we can easily tell whether intelligent machines can have humanlike emotions. My guess is that they can. They will, of course, be machine emotions and not *exactly* human emotions. But if, in the future, an artificial neural network brain performs in a similar way to a human brain, then presumably the artificial brain will have emotions and feelings in a similar way to the human brain. We can tell, however, that machines can certainly have behaviours which can be regarded as emotional, in the sense that if exactly those behaviours were exhibited by a human or animal, they would be regarded as being emotional ones.

The Brain Builder group in Kyoto, Japan, has claimed that it will have a machine with intelligence similar to that of a cat by the year 2001, although one does not expect it to behave *exactly* like a cat. For example, it will not need to drink milk. Nevertheless, I think we can expect such a machine to exhibit clear catlike emotions and to exhibit self-will, in that a cat usually does what it wants to do and not what a human wants it to do. Maybe the group will achieve their catlike brain, and maybe not. It is important that they believe it to be possible.

A critical point is that because machine brains and human or animal brains are different physically, so their characteristics are likely to remain, by and large, different. As discussed previously, machine intelligence is different from human intelligence. In some ways it is better and in some ways worse. But other characteristics, such as creativity, learning, self-will and emotion will also be different, to an extent at least. With emotions, humans have their own set of acceptable and expected feelings and responses. We cannot, however, expect machine emotions to have *exactly* the same values as human emotions.

When another human dies, particularly one with whom we have had a long association, we will probably be extremely upset and may cry. In particular, if the other person dies when still young or in an unfortunate way it can be especially upsetting. Even when a human we do not know dies, in certain circumstances we can feel strong emotions. On the other hand, if a bee dies this can make us feel happy if it has been annoying us. Or if an animal dies in front of us, it may disturb us immediately, but if we look away any emotional upset will soon quickly pass.

Now let us consider a machine and assume that it does have emotions. If so, given that some of these may be due to its initial program and some due to what it has learned from experience, what will its emotions be like? This is very difficult to imagine, but try to put yourself in the position of the machine and see what emotions you might have. Electrical power would probably be important, and all things associated with it. Animals might not bother us too much, but might upset us if they affect our senses. Other machines, with which we cooperate, could affect us positively. However, humans trying to switch us off, or break us up, could be particularly annoying and could easily make us angry.

Humans usually show little or no positive emotion towards machines, but we often get annoyed with them and kick them or hit them when they do not do what we want. Indeed, when a machine no longer works well, we simply throw it away. However, when a machine looks cute, is small, does not harm us and behaves in an animallike way, then we can start to feel emotionally positive towards it. On the other hand, if a machine looks dangerous, is large, is likely to harm us and does not behave in an animallike way, then we are likely to feel scared or frightened of it.

Because human brains and machine brains are different, machines will exhibit different forms of emotion, self-will, consciousness and so on, from humans. If we try to make a machine brain as close as possible to a human brain, perhaps using biological techniques, then it is likely that the machine characteristics will become similar to human characteristics, only being *exactly* the same if the machine brain was, in fact, a human brain. Even a biological brain with fewer neurons, connected up differently, would exhibit different characteristics, as is shown by cats, dogs and bees.

Maybe a machine will not cry, but maybe it cannot and maybe it does not need to. Maybe a machine does not scream when it is attacked, but maybe it cannot and maybe it does not need to. But what if a machine's circuitry gets hotter when it is attacked? What if it sends out very high-frequency signals? What if it moves away quickly from the predator? Some will say to this, 'But this is not emotion, is it? This is not self-preservation, self-will, awareness of others or understanding danger. How can it be, this is only a machine!'

In the same way, we humans can say, 'The machine is not as intelligent as a human', or even, 'Machines are not intelligent at all.' As with other characteristics, such as emotion or self-will, we can then find, we hope, some definition of intelligence which allows us to make statements indicating just how much better humans are than machines. But as time passes, it gets more and more difficult to find aspects of intelligence which we can use to show our human superiority. Consciousness is one of the most recent of these. To be truly intelligent, some people might say, a machine must be conscious, like a human or animal, and we can clearly see that it is not.

Definitions of intelligence now seem to last only two or three years before being made obsolete when some new feature is incorporated in a machine, often in a much superior way than in any human. As each year goes by, machine brains are becoming more powerful, more efficient and exhibiting ever better performance in the areas of intelligence in which machine brains are already superior. They are also rapidly closing the gap in the areas of intelligence in which machine brains are inferior to human brains.

An artificially intelligent machine brain is certainly physically different from a human or animal brain, not only because it is usually electronic or optical, as opposed to biological, but also because, at present, an artificial brain is usually smaller, containing perhaps only 50 to 1,000 neurons compared with a human brain's 100 billion. Only when artificial brains start to contain similar numbers of neurons to human brains can we then get emotions and other characteristics which are similar to those of a human in *all or most* respects.

But why can we not yet have artificial brains with thousands, millions or even billions of neurons? There are two main problems. Firstly, the ways in which artificial neural networks are taught are still not fully developed, particularly when many neurons are involved, although research in this area is progressing rapidly. Secondly, artificial neural networks at present usually have only 20 or fewer connections, i.e. synapses, between neurons, whereas human brain cells have many more. Indeed, some have 70,000 or even 80,000 connections.

However, machine brains are already much faster than human brains. In the time it takes a human neuron to change its response (something it can do about 1,000 times a second), an electronic transistor can typically change its response a million times; that is to say, an electronic neuron is a million times faster than a human neuron. Furthermore, each year this machine superiority increases.

Electronic computing machines are generally extremely accurate and have high precision. Human neurons, however, tend not to be so accurate or to exhibit such precision, although collectively, human neurons do seem to be superior to other animal neurons in this respect.

The same comparisons can be made with many features. Human neurons seem to work better than other animal neurons in terms of mathematics, memory, abstract thought, and so on, yet in each case artificial machine brains can perform better than the human version. So in many respects, just as human brains are superior to the brains of other mammals, so machine brains can be superior to human brains. At present this machine superiority is limited to a range of characteristics. A key question is: 'How broad can this superiority be?'

The synapse connections, the wiring, of a human brain can appear to be fairly random. It certainly seems so when the brain is minutely examined. My own belief is that certainly the wiring is extremely complex and difficult to comprehend, but it is affected by our genes. It is directly a part of our initial program, our human genetic inheritance. The different synapse operating conditions are then learned as we experience life, and as our brain grows biologically.

So, as humans we have a number of basic instincts, caused by the initial physical arrangement of our brains. These instincts are mostly shared by all humans. However, because of our genes, some particular aspects are initially more strongly linked, some more weakly. Different humans, therefore, have different natural aptitudes. Our subsequent learning and experience merely builds on this.

Such 'random-looking' human neuron wiring connections have arisen through genetic breeding over many years, with slight modifications and different synapse strengths based on relatively small changes at different stages. However, this seemingly random arrangement of a human brain should not be seen as something better than a typically more precise and structured machine brain. Indeed, it could be argued that better operation, better maintenance and fault correction can be obtained from a more precise, structured approach. Certainly when something goes wrong in a human brain, it is often difficult or impossible to do anything constructive about it because of our lack of understanding. The gene-dependent human brain arrangement does, however, give rise to distinct characteristic differences between individual humans. If, on the other hand, all brains were structured in exactly the same way, differences would be much less and would essentially be based on each individual's experiences and physical differences.

Tidying up the operation of a brain, in a machine brain, does offer a lot of possibilities in terms of functionality. In consequence, because human brains appear as they do now does not mean that they cannot be improved upon and even be completely surpassed by some other, more functional, brain.

At present the density of neurons in a human brain – that is, the total number in the space provided – is still considerably more than in usual computing machines. However, whilst the human neuron total remains fairly constant over time, the machine neuron total and density is increasing rapidly. A present-day Intel Pentium chip houses more than three million transistors in a space the size of a pea. It is widely anticipated that only a few years into the next century machine brains will have surpassed human brains in processing power, in the same space occupied by a human brain. That is with available technology, as we know it today. With one or two technical breakthroughs the advances could be even quicker. Optical devices such as laser communication, for example, offer enormous possibilities in terms of replacing conventional machine brain wiring, thereby giving rise to a major shift in available power, size and speed.

It is very easy to look ahead and see clearly that machine brains, which are already superior to human brains in many ways, will become far superior in many other ways in the next decade or so. Also, machine brains can be custom-designed and can conceivably be made as large and as powerful as required.

172

Human brains, which grow from a single cell, will remain the size they are, nothing more, nothing less. We will look further at this in later chapters.

As machine brains are going to be superior to human brains in many ways in the not-too-distant future, are there any ways in which human brains are going to stay ahead, possibly for ever, or is it just a matter of time before machine brains completely surpass those of humans? Let us look at the defensive arguments which some people use in supporting the human case.

Firstly, it should be remembered that both human and machine brains are simply physical entities, albeit extremely complex and difficult to understand. Secondly, machine brains and human brains are different: both physically, in terms of what they are made of and how they operate; and mentally, in terms of their capabilities. It is not correct, therefore, to say that because a machine brain will never be *exactly* the same as a human brain, it will always be inferior to it.

Pro-human argument 1: There is something magical about a human brain, a secret ingredient, a characteristic which cannot be physically explained. Furthermore, without this magical ingredient a machine brain can never perform in the same way as a human brain and can certainly never be superior to it.

I do not attach much importance to this argument. Simply describing as magical something that we do not yet understand physically is merely a way of avoiding the issue. Also, if we assume that there is a magical ingredient and we do not know what it is, how can we know that it is also the important ingredient that a machine brain needs in order to be superior to a human brain? Indeed, it may well be that a machine brain can have not only this magical ingredient but many more of its own. As we do not know what the magical ingredient is, there is no way that it can be logically concluded from this that a machine brain cannot be superior to a human brain. Hence, pro-human argument 1 does not hold water.

Pro-human argument 2: We still cannot model the physical and operational aspects of a human brain in sufficient depth to enable the exact working of the brain to be copied by a machine brain. Possibly we need to look very deeply to quantum theory in order to describe the operation of a brain accurately. This is the Penrose argument.[34]

I believe this argument, as written here, to be reasonably sound. However, even if it is true, we can probably get fairly close to making a machine brain operate like a human brain without it being *exactly* the same. My argument here is similar to that I used about pro-human argument 1: as we do not know all we need to know about a human brain's operation, how can we be sure it will be significant when we do discover it? We certainly cannot conclude from this that because a machine brain cannot be *exactly* the same as a human brain, it cannot be superior to a human brain.

In both of the pro-human arguments so far, it is necessary to take an illogical step in order to prove a point. This step is as follows. Statement: A machine brain cannot be exactly the same as a human brain. Conclusion: A machine brain cannot be superior to a human brain.

Such a conclusion is flawed. Just because something cannot be exactly the same as something else does not mean that it cannot be better than something else. To stress the point, consider this statement relating to American football teams: Pittsburgh Steelers cannot be exactly the same as Dallas Cowboys: Conclusion: Pittsburgh Steelers cannot, therefore, beat Dallas Cowboys. Obviously this is nonsense.

Pro-human argument 3: There are certain characteristics of a human brain, perhaps consciousness, which, given our knowledge at the present time, simply cannot be re-created in a machine brain.

This is, in many ways, a similar defensive statement to the Penrose argument, although it comes from a different angle. In the same way, incorrect conclusions can be drawn; for example: because a machine brain cannot be conscious, it cannot be superior to a human brain.

As before, the missing link to the operation of a human brain, in this case consciousness, is something that we do not really fully understand at present, although, as with the quantum theory approach, some people feel that maybe in time we will indeed get to understand it better. This latter point then raises a further issue, that as and when we do fully understand human consciousness and/or quantum human brain operation, and we can physically model it, then and only then can machine brains be roughly equal to human brains. But, because they also have many distinctive advantages, at that time machine brains will in fact be superior to human brains.

So, interestingly, pro-human arguments 2 and 3 can both be taken to indicate a future in which human brains can be surpassed by machine brains, once we have grasped a better understanding of how a human brain works. This could be in a hundred years' time or perhaps in only a month's time. But who says this is the case? Is it the medical doctors, neurosurgeons or psychiatrists who work with human brains every day, or is it the engineers who are building machine brains? It is usually none of these, rather it is those who wish to think about the problem for some time to come, receiving support for related research and having numerous academic papers published on the topic at conferences. There is usually a very strong self-interest for those who believe consciousness to be a major stumbling block.

All the major pro-human arguments can, I feel, be drawn into at least one of the three given here. Indeed, I find myself agreeing with argument 2 and possibly 3, although certainly not with some of the conclusions that have been drawn and certainly not with argument 1 at all.

A cow's brain is different from a human brain, and so a cow is conscious in a different way from a human. A cow is aware of other cows and, in its way, is aware of itself. A cow exhibits fear or pleasure or self-will in its own way. The two brain types are essentially different and we know that, by and large, a human brain usually performs in a superior way to a cow's brain. The same type of argument can be applied to all mammals and creatures of the earth. Similarly, machine brains are different from human brains. A machine brain can be conscious in a different way from a human brain. A machine can be aware of other machines and a machine can exhibit fear or pleasure or self-will in its own way. The two brain types are essentially different. Here I am, of course, referring to a machine which is permitted such characteristics – a tabletop personal computer is not usually allowed to function in this way.

But one cow is not very different from any other cow and one human is not very different from another human. In both cases they are members of a particular species. Machines, however, can be, and indeed are, very different from each other, carrying out varied physical actions with greatly differing machine brain activities. With intelligent machines we are not looking simply for the addition of one extra species, but rather a whole range of species, a range of new life forms.

There are those who believe that humans, mammals and machines are all different forms of one common being. For example, in his book *Chance and Necessity*,[47] Monod said, 'The cell is a machine. The animal is a machine. Man is a machine.' Simons[48] wrote, 'Man is a mechanism, a robot, programmed for performance, but a robot with remarkable insights, talents and emotional sensitivities.' Thaler,[46] until mid-1995 a research scientist with McDonnell Douglas, put it another way: 'People will start to ask the question, "How am I distinct from these machines?", and I think the inevitable answer is that we are the same.'

Humans, then, can be regarded as simply one type of machine, a biological, electrochemical form. In looking at it in this way, it can be seen that humans are simply one model, one way of doing things, one approach which has arisen naturally. We think and act in our way, and other machines will think and act in other ways.

Although I can see the point in reflecting on humans as one type of machine, I wish, for the purposes of this book, to keep them separate. Machines themselves provide an enormous variety of possibilities, from very simple mechanical switches to extremely complex artificial neural networks. The range of human capabilities is also large, but not to the same extent. So here, we will consider humans and machines to be different, because my definition of what is and is not a machine is perhaps more restrictive.

One thing worth pointing out, however, is the following. With some particular machines we know about, we are quite prepared to accept that there will soon be something much faster, which can carry out its function more quickly, and which can deal with more things at the same time. We also accept that the better machine will itself, in time, become surpassed and obsolete. We humans do not like to look at ourselves in the same way, and yet we should.

By considering consciousness or quantum theory we can look at one aspect or another of a human brain's operation and conclude that a machine brain will not perform in exactly the same way. This is because either it is not conscious in exactly the same way as a human brain, or such a machine brain cannot be constructed because we do not yet know exactly how a human brain works. Neither of these features prevents a machine brain from performing in a superior way to a human brain.

Only those who think that a human brain has some magical element appear to believe that machine brains may never surpass human brains in performance. Both the quantum theory viewpoint and the consciousness viewpoint appear to indicate that there merely remains one step in understanding and we are there. My own belief is that whether or not that step exists does not really matter. We can already build artificial neural networks, without resorting to quantum theory and with a consciousness different from that in humans, which exhibit a range of behaviours and perform, in many ways, better than human brains. In the future they will exhibit a more powerful performance and will, before too long, surpass the overall performance of a human brain.

The performance of machine brains is undoubtedly getting closer and closer to that of human brains, in those ways in which they still lag behind, and as technology improves, so machine brains will benefit. The overall performance involves a wide range of characteristics, with intelligence playing a key role. As time progresses, so the intelligence of some machines will, I believe, become roughly equal to that of some humans. Once we are in this state then it is apparent that machine brains will get more and more intelligent, while human brains stand still.

The one problem which remains is scaling up. We now have artificial neural networks which operate with 50 or 1,000 neurons, but to compete with human brains these networks must contain billions of neurons, with partitioning and different functional operations, in order to cope with different aspects of performance.

This scaling up, involving connecting more and more neurons together, with its inherent problems of getting the networks to learn and function in what is seen as a coherent way, presents the biggest hurdle. Subsequently such machine brains can have tens and then hundreds of billions of neurons and hence will be able to outperform human brains.

A major problem in the past was to realise machine brains which could learn, be creative and aware and exhibit free will. I, like others such as Thaler,[46] believe these features to be in place now, and that it is only the scaling up that remains. Indeed, many of these features are well demonstrated in the robot machines described in the next three chapters. A key fact is that for both humans and machines, our mental state is dependent on nothing more than our original gene program and our subsequent experiences.

So, is it sensible to say that machine brains can one day be more powerful than, more intelligent than and superior to human brains? All the physical indications are that this will be the case. In terms of computing power, in terms of speed of operation, memory capabilities, learning, creativity and so on, all the indications are that the gap between human and machine intelligence is closing at a fairly rapid rate. No theory or physical evidence, barring magic, stands in the way of machine intelligence firstly becoming roughly equal to and then surpassing human intelligence. True, machine intelligence is unlikely to be achievable in exactly the same form as human intelligence, but that is a different issue and the two should not be confused. I certainly believe that it is possible for machine intelligence to surpass human intelligence, because the scientific evidence points that way and there is nothing that leads us to say 'No, it cannot happen.'

This means that such intelligent machines will no longer simply help humans to use our own intelligence, but will be intelligent in their own right. These machines could then be used as authoritative voices, could be asked for advice on different topics or could, very swiftly, solve a lot of the world's human-designed problems. They could design new means of travel, create soft drinks, invent musical scores or discover new materials, as does Thaler's 'Creativity Machine',[46] which is essentially a neural network housed in a computer workstation – a stand-alone machine brain. The network takes, as its input, information in the form of 'off' or 'on' states and gives, as its overall output, another set of 'off' or 'on' states. Interestingly, certain neurons in the middle of the network appear to get more excited when certain features appear at the input.

When will all this be? Realistically it is anybody's guess. Some say a hundred years or so, or maybe even several centuries. However, a realistic look at the rate of change of technology and our rapid progress with machine intelligence makes these appear to be extremely conservative views. Perhaps it is a rearguard comforting action for humans, however, in that if we humans must concede that machine intelligence will one day surpass human intelligence, something we naturally do not like, then let us put it off for as long as we can, certainly for a century or so, and hope that it goes away!

Some scientists have argued that in only ten or twenty years' time, machine equivalence will have been gained. A reasoned case, indicated by Moravec,[25] based on the accelerating rate at which technology is changing, and what has already been achieved in machine simulations, points to this being in about the year 2030.

Indeed, the increased power, speed and performance of a machine much reduces the scaling–up problem, which is all that remains, so that a machine brain with 100 million neurons may well be able to perform in a roughly comparable way to a human brain with 100 billion neurons.

It is also worth remembering that human brains tend to do many 'unnecessary' or unprofitable things that a machine, or other life form, does not need to do, such as worrying or dreaming during sleep. Also, large portions of the brain deal with human physical characteristics and senses – for example, moving hands and seeing – and these may not be relevant to a machine. Clearly, if some of these things are avoidable it may be that a slimmed-down machine network version can perform in much the same way as its larger human counterpart. In fact, many human values and interpersonal skills may not be necessary for a machine form.

It is surprising how humans, for the most part, appear to accept as a basic principle that they are the superior beings on Earth, that is how it is now and that is how it always will be. The concept mentioned earlier that the human brain has some 'magic' component which keeps it ahead of the field is based on this principle. So too is the conclusion that if a machine brain cannot achieve *all* the features of a human brain; then the machine will always remain subservient to humans, no matter how far it advances with regard to speed, capacity and logical design. This is clearly wishful thinking on the part of humans.

Another possibility, as indicated by Moravec,[25] is that we humans will be able to switch our 'mind programs' into the plastic or shiny metallic bodies of our choice. In this way we would be able to give ourselves a type of immortality, with not only our body being replaced by a new gleaming robot body, but also our brain being replaced by a new, much more powerful, faster, more accurate machine brain. The concept is that our own brain pattern at a particular time can be transferred across into a machine brain and start a new life there, whilst retaining all memories and experiences up to that point. In other words, a human is born with an original gene-programmed brain which is then trained in terms of human experiences. Eventually the status and arrangement of the brain is switched over in its entirety, to act as a start-up program for a new machine brain in which further learning and experiences progress. Indeed, as it is a brain pattern that is being transferred, there is no reason why the original human brain could not go on operating in exactly the same way, with the machine brain acting as a clone.

All of this requires a much deeper understanding of how human brains work and relies on the idea that we would be able to open up a human brain, and make a full, detailed record of its status, synapse connections and arrangement, without necessarily affecting the brain detrimentally. In reality, at the present time we are nowhere near such a position. In addition, the overall concept is that the human brain and machine brains will be directly compatible, whereas in truth this is also a long way from fulfilment.

Effectively the machine brain would be simulating, or copying, the operation of a human brain, and would be able to do much more because of its improved power and capacity. The end conclusion of this is that a new life form of machines would appear, whose intelligence originates from human intelligence, although the actual level of intelligence would very quickly far surpass anything that a human could achieve.

I am extremely sceptical about the realistic possibilities of directly transferring human brain patterns in this way, partly due to the apparent problems in getting inside a living human brain without destroying parts of it, but much more so because the differences between machine and human brains mean that a particular human brain pattern would not necessarily fit snugly into a machine form.

However, it is quite conceivable for certain human characteristics to be passed on by a human into a machine brain. This could be done by programming the machine form on start-up, or by teaching the machine to operate in the same way that the human does. This latter concept is the same type of approach as creating an expert system.

Transferring our human brain patterns across to a machine brain does, however, open up all sorts of interesting possibilities. In the first instance, as the capacity of the machine brain could be much greater than that of a human brain, several brain patterns, from different people, could be transferred into one machine brain, possibly leading to a multi-personality brain. Husband-and-wife brain patterns could be transferred to reside together in the same machine brain. They would then be able to communicate, internally, directly with each other. Indeed, several husband-and-wife combinations could reside in the same machine brain, leading to mental partner-swapping and possibly to substantial internal arguments. It would be rather difficult to leave your partner, though, if you fell out.

It is a trait with humans that we somehow wish for the human race to stay in charge. We would like to find a way in which we humans will remain in control. Even if machine brains do become more intelligent than human brains, then we still feel that we should end up in a commanding position. In thinking of a mind program transfer we also probably conceive of a shiny,

181

metallic humanoid robot form, having a powerful machine brain. It is then simply a case of transferring our human brain set-up into the robot, with little or no disruption to ourselves.

As a challenge to this view, the first machines which could be considered as more intelligent than humans could easily be large computers which have been given the capability to learn about and control financial transactions, with no need to move around or to use visual or audio senses. This could easily happen in the next ten years or so. Life for an ex-human brain, inside a box which deals solely with non-qualitative financial transactions, would then not be seen in such a positive way. The basic starting point is that machines do many things differently from humans and have separate, often more firmly directed, goals. Furthermore, it cannot be expected that even robot machines will do all the things humans do. Indeed, why should they? After all, they are machines, not humans.

Another point is that once the first machine has appeared with an intelligence superior to that of a human, it is extremely unlikely that that machine will be happy to have its own machine brain pattern wiped clean, only to be replaced with a less intelligent brain pattern from a human. In particular the original, more intelligent, machine brain will probably be very well suited to its particular situation, will be quite happy residing in its own machine and would not wish to be downgraded by the imposition of a human brain pattern. Essentially, human brains and their associated brain functioning are suited to the human body in which they reside, or possibly another human body, but life in a machine brain is a very different proposition, particularly if that machine does not move or function like a human.

Machines and humans are different. Machine brains and human brains are different. When machine intelligence surpasses human intelligence, this will be in terms of machines *per se*, in their own forms, shapes and sizes. However, as they are more intelligent, very soon it will be they that decide how they operate, what they do or don't do, when they switch on or wake up and when they stop work. After all, they will be more intelligent than us. They will be able to make the important decisions and stay one step ahead of us.

As discussed in this chapter, the only things which appear to stand in the way of machines being more intelligent than humans, excluding the magic theory, are either our lack of a full physical understanding of how a human brain works, which quantum theory may solve, or the consciousness which humans have. The first of these, the physical understanding, is, however, merely a source of delay, in that once we gain a final complete understanding, machines can then be more intelligent than humans! So, is consciousness in a human brain the one saving feature of the human race?

Personally, I do not feel that human consciousness is a stumbling block to machine intelligence. There is no reason whatsoever why machines should be conscious in the same way as humans. Intelligent machines of certain types may indeed be conscious in a machinelike way. After all, they are based on signals and messages in a similar way to a human brain. But we humans must look at a machine brain from the outside in. Only if we ourselves had machine brains would we be able to say whether or not they were conscious in the same way that we are. But then we would not really be humans, would we? We would be machines.

It is very difficult for us to say exactly what human consciousness is, although an attempt was made towards the end of Chapter 4 to bring a number of ideas together. We know it is there. We know that we, as humans, are conscious, but we may never be able to describe what consciousness is exactly. We would perhaps like to think that machines are not conscious. Realistically, maybe they are, in something like a human way, but maybe they are not. It is difficult to say there is nothing there in the case of an intelligent machine, although it is certainly not yet the same as a human. It is rather like another animal or an insect.

But consciousness is just one characteristic of the functioning of a human brain, and we know that there are many others. Many of these features can be, and indeed are, much less abstract and possible to measure. Of those we can measure, either we know already that machines are much better than humans, as with memory and speed, or we can see that logically they will be better before very long, as with the ratio of power to size.

A direct analogy can be drawn with a soccer match between

humans and machines. The human team, let's call it Man United, is generally poor but has Consciousness, considered by some to be the world's top player, on its side. The opposition, meanwhile, let's call it Bayern Machine, has some good players such as Memory, Speed and Mathematics. Indeed, they have just brought in a new player, Creativity. Having the world's top player on their side does not mean that humans will win the match, or does it?

Bayern Machine is frequently recruiting new and better players. Indeed, many transfer across from Man United. The opposition is constantly getting stronger. The humans still have Consciousness, though, and some feel he will always be our best player, unless we can again harness machine power for human benefit by using machines to provide humans with Super-Consciousness or Hyper-Consciousness.

In the world soccer match there is no final whistle, or rather it is so far into the future that it is irrelevant, so we cannot hold on to our lead until the year 2010 and then relax because we have won. The game goes on, and keeps going on, and all the time the machines team is getting better and stronger.

At present Man United is winning by something like 5 goals to 3. But we have, unless we can sign up Hyper-Consciousness, scored all the goals we will ever score. Bayern Machine, meanwhile, will continue to score goals until they are in the lead. Then, on the assumption that the human team does not recruit any new players, whereas the opposition can continue to strengthen its team, the likelihood of Man United regaining the lead, even momentarily, is negligible. Once Bayern Machine are in the lead, they will probably stay there.

We should not and cannot say that we have Consciousness in our team, therefore we will win and we do not need to worry about the opposition. Bayern Machine will certainly not say, 'OK, Man United have Consciousness on their side. Let's not bother even turning up to play the game.' Consciousness is just one player with some good points and some bad, and indeed he may not be a match-winner at all. The machine team may well find itself a player who is much better than him. Perhaps he will go under the name of Machine Consciousness. They have already found better players than most of the players in the human team, so why not with Consciousness? If that happens, humans appear to have no chance. Even if the Bayern Machine team does not come up with such a star, they still have an extremely strong team which is getting stronger all the time. How long can Man United hold on before the game is lost?

The analogy with a soccer match was drawn to show that consciousness is just one feature of a human brain's operation. It may or may not be critical in terms of machine intelligence surpassing that of humans. Indeed, it may be of no consequence whatsoever, but assuming for the moment that it is an important factor, it is left standing as the final obstacle for machines to overcome. Just like the quantum theory suggestion, consciousness is providing a standard to be surpassed in time. Whichever way we look at it, there is only one conclusion: it is only a matter of time before there is nothing humans can do to stop machines becoming more intelligent than humans.

Chapter Nine

The Reading Robots – An Overture

There is an expression, 'The proof of the pudding is in the eating.' I am not sure that it is exactly the one I am looking for here, but it will have to do. The discussion in the last few chapters has been largely of a philosophical or theoretical nature, and as far as this goes a range of conclusions can be drawn, based on well-reasoned arguments. I am clear about my own opinions but you, like all other humans, will probably have your own opinions. If they are good, sensible opinions they will certainly lie very close to my own.

But no matter how much philosophising or theorising goes on, the most powerful test is in terms of real, practical machines and their intelligent characteristics compared with those of a human.

The roots of this book are based on the firm foundations of practical experience and experimentation. The conclusions drawn in previous chapters relate back, and are dependent upon, those firm footings. Even the look ahead at the start of Chapter 2 depends on a firm, practical starting point.

In this chapter, and the next two, we look at some of the practical work which has gone on in the last few years in the Department of Cybernetics at the University of Reading. These chapters, for the first time, bring some of this work together as a cohesive whole, and provide strong supporting evidence for this book. All the practical work described here has actually been, or is being, carried out, and the robot machines described actually exist. In one or two cases I refer only to computer simulations. However, for the most part I describe practical physical systems. In each case these actually work in the ways described.

Just about any experiment can be carried out in terms of a computer simulation. The trouble is that virtually any results the experimenter wants can be obtained under certain circumstances. The computer simulation world is an ideal, a type of perfect, unreal world in which things can be shown to happen or not to happen, largely as the experimenter desires. The real usefulness of computer simulations is in testing out theories before trying out the real thing, or in carrying out designs, or in designing computer systems to operate interactively with the real world. The key element comes in the actual link with the real world.

If you are told by someone that he has shown conclusively that machines are, or are not, more intelligent than humans, or he has designed 600 robots which sing and dance and tell jokes in Russian, then by all means question him. Has he done this as a purely theoretical exercise, or computer simulation, or does he actually have the machines which are intelligent and can sing and dance? Unless he can produce the goods, his results are of little or no consequence. An actual robot walking machine which takes one step and then falls over is worth far more than a simulation of 29,000 robots running the London Marathon in record time.

It is important, too, to take account of the wide range of capabilities which intelligent machines have. It can be easy to fall into the trap of thinking about intelligent machines simply as computer workstations which sit on a table in front of a person and only do something when a letter or number is pressed on the keyboard. It is very difficult in this case to imagine a machine understanding complicated oral messages, and it is easy to regard the machine as being stupid because it does not really understand a human language, such as English. It can also be difficult to think of the computer as being something which can learn, be creative and express emotions.

In 1995, at a scientific meeting, I was approached by a computer scientist, a senior academic, who was convinced that machines could not learn but could merely change their programmed responses, in a programmed way, to external stimuli. So I invited him to Reading to see our robots, and to his great credit he came. He left on the evening of his visit a converted man, having witnessed at first hand some of our robots learning.

It transpired that he had, until that time, dealt only with tabletop computer workstations and had little or no idea of what was possible otherwise.

Robots which can learn are just one part of the work taking place at Reading. The total repertoire ranges from computer packages operating on large machines, through computer simulations (yes, even we do some of these!) and special-purpose technical solutions to robot machines. To describe the entire set-up would take too long, so here I mention only the projects of particular relevance, to serve as examples rather than give the complete picture. An overriding priority of the work is, in each case, the desire to try out a new idea, actually to implement it and to see how it works in practice in the real world.

Let us start at the big computer work, the sort of thing which was discussed in Chapter 6. Here, a range of intelligent techniques is being employed to monitor and control machinery, to oversee and analyse the operation of UK national utilities and to detect and diagnose faults in large-scale manufacturing systems. Let us look at an example to show what is done.

With high-speed, high-performance production machinery, the drive is constantly on to achieve machinery which is fast and efficient, requires no human intervention when in operation, does not break down, can be operated constantly and can itself decide what maintenance is required and inform well in advance when work is necessary. Let us assume for the moment that the machinery is producing beefburgers,* packaging them and getting them ready for transportation. The quicker the machinery can do this, at the lowest possible cost with as few breakdowns as possible, the more efficient is the operation and the higher the profits.

At many different points on the production line, one can obtain information such as the speed of a conveyor belt, the rate of beef mixture arriving, the quality of package sealing and so on. All these different factors add up to indicate how well the production process is going. Possibly 20 or more different factors are relevant and measurable at any time. Some combinations of

---

* For reasons of commercial confidentiality I do not mention the actual products with which we work.

these factors indicate that the machinery works well, while other combinations reveal modes in which a problem is likely to occur, possibly resulting in the machinery having to be stopped.

With our human brains we can deal with two or three factors at the same time, and adjust, for example, the speed of a roller, the size of burgers and the moisture content, in order to achieve good production. However, where 20 or more factors are all interdependent on each other, it is simply not possible for humans to cope. Different methods have therefore been devised to teach a computer-based system ways in which the machine operates well and ways in which it operates badly.

One procedure is the use of an artificial neural network which learns nonlinear relationships between the different factors. A human can teach it to recognise when the machinery is working well and it learns by experience which combinations of factors achieve this. It can also be told when the machinery fails and how it has failed, and thereby it learns which combination of factors leads to the failure. Failures can often be caused by an extremely complex relationship between the factors, which a human operator would probably not be able to recognise.

By monitoring the machinery for some time, the artificial neural network is able to obtain a detailed picture of the process. The network has essentially learned how the process works.

Subsequently, the network can monitor the production process in operation. If the measured factors are typical of normal operation then a positive indication can be given. If, however, the network discovers that some of the factors are drifting towards a combination which is known to produce problems, it can indicate that this is so and the requisite adjustments can be made, either automatically or by a human operator. In this way the network can direct machine operation in a way that a human could not.

In the case described, an artificial neural network can learn the best mode of operation for an individual production line. The network is, therefore, uniquely trained to that line. Different human operators, meanwhile, tend to vary distinctly in their operation of different lines, and may even vary considerably in their operation of the same line. The use of Reading software on certain production machinery in the United States revealed considerable differences in how different human shifts were operating, so much so that considerably more machine down-time was being brought about by one group as opposed to another.

The work described, as one might expect, largely makes use of tabletop computers, the overall aim being to enhance standard machinery with limited intelligence.

Another example of this is the use of an artificial neural network to identify features on human faces. The driving need is for the image of a moving face to be sent down a standard telephone line in real time, to accompany a conversation using videophones. To send the whole image of a face down a line leads to enormous problems simply because of the need for expensive lines to transmit such a volume of information. It would be preferable to pick out certain features, such as mouth and eyes, and to send a lot of information on these and very little on other features. This is good as far as it goes but faces are all different, so how does the transmitting system know where to find the eyes, nose or mouth on a face it has never seen before?

An answer is to use an artificial neural network which is trained to recognise certain features on the image of a face. For example, the network is shown images of many different faces and is taught to recognise an eye. When it then sees a new face for the first time it can very quickly pick out the eyes in the image.

The work on this at Reading mainly involved Mark Bishop, who is now a lecturer in the department. At first the results did not seem encouraging, until it was realised that the predominantly male researchers 'helping' with the project had taken the neural network to the university students' union building and trained it to recognise the eyes of students who, for some reason, were mostly female! When it was returned to the department, the network then performed poorly at recognising male eyes. However, once the network was trained on a reasonable mixture of eyes it performed much better, although it still occasionally picks out a heavily dimpled chin, Kirk Douglas-style, on a face it has never seen before, and considers it to be an eye!

One computer simulation project worth mentioning is that carried out by my former research student, Nigel Ball, who is now at Cambridge University. In this work an artificial organism, a simulated life form within a computer, was created and was given a world to live in, consisting of a maze with numerous stop-off points. At some of the stop-off points water is placed, and at others food. The goal of the organism is to move around the maze, to learn where the food and water are located and to return to them periodically to 'eat' or 'drink'. As its food and water supplies are used up, so the organism has to move around to stay alive, collecting more food and water.

Just to make life interesting for the organism, 'nasties' can be placed in the world to chase it around, and the supplies can be moved, so the organism must learn their new locations. Interestingly enough, the organism, which is based on a complex bank of artificial neural networks for mapping, learning and association, tends to return to old supply locations even when the supplies have been moved elsewhere. It learns fairly quickly about new locations but only gradually forgets about former ones.

One extremely rewarding area of research in the department is the design of technical aids for people with disabilities. One project a couple of years ago was to build an electric-powered platform for wheelchairs, which can be used by children who otherwise must use a manual wheelchair. A manual wheelchair, with child on board, is pushed up a ramp on to the platform and clamped in place. The child can then, by means of a joystick, drive the platform around. The platform also has ultrasonic sensors, which send out high-frequency sound signals to the front and sides. If something such as a wall is fairly near a sensor, then a sound signal is reflected and received by the sensor. The platform is fitted with a buzzer device and can also make use of a voice command. When an obstacle is detected, the buzzer sounds or the voice says, 'Object to the right', or, 'Object to the left', as appropriate, whilst the platform is automatically prevented from actually hitting the obstruction.

The electric platform was put together for the Avenue School, Reading, a school for children with special needs, and operates there on a regular basis. Essentially by learning to use it, children are learning how to use their motor skills to operate an electric-powered wheelchair. The same ultrasonic sensors are also employed on the Seven Dwarf research robots, described later in the chapter, showing how results and work on one project map over into other related research.

The electric platform was demonstrated on BBC TV's *Jim'll Fix It* programme, in which Sir Jimmy Savile made dreams come true for children and adults alike. The cybernetics department had a number of links with this programme, and Sir Jim in particular. More of that shortly.

We had one surprising result from the electric platform when we first drove it around to try it out. Together with the students who actually built it, I found it difficult to control and awkward to steer. We were all used to driving automobiles but this was quite different. We were therefore worried about how the children from the Avenue School would find it. Perhaps it would be too difficult for them. We need not have worried. The first child to try it out, Ranjit, immediately started driving the platform around at a frightening speed, being able to swerve around objects and corners. Although he has cerebral palsy, Ranjit was easily able to outperform all of us. Our team included a student whose hobby was rallying, but even he could not compete with Ranjit. But Ranjit was by no means unique, as most of the children could drive the platform around in a similar hair-raising fashion.

Projects such as this are extremely rewarding when successful results are achieved. It is something which we, as a department, can do to help people with disabilities in a way that companies would perhaps not, because there is no commercial gain for them. However, such projects are also eye-opening when it comes to learning about human intelligence and how machines can help to allow a human better to use that intelligence. As discussed in Chapter 7, with the example of the autistic person in *Rain Man*, certain people are just as intelligent as, or perhaps far more intelligent than, 'Mr and Mrs Typical' but are often regarded as being inferior because they are physically or mentally different in some way. There is much, much more we could usefully do for people with disabilities, by designing machine technology to assist them.

Other such projects have included a telephone system for people who are deaf, enabling them to use sign language input through a camera rather than a standard telephone handset. Matchstick-like hands then appear on an ordinary television screen at the other end of the telephone line, moving in real-time response to sign language input. Although restricted in the overall signing vocabulary, the system does allow a real-time signing conversation to occur.

Another project recently completed by students was a clever bath for people who have epileptic fits. It is obviously better for a person who has epilepsy to be able to bath, and indeed do other things, without someone watching over them all the time. The bath had to detect when a person had a fit coming on, so that water could immediately be pumped away at high speed to avoid possible drowning. Momentum for the project was achieved when a young local woman sadly died in such circumstances. During the project a range of solutions was looked at, whilst attempting to avoid covering the person's body in wires and sensors.

When an epileptic fit occurs, although the person has much abnormal brain activity, there are not necessarily many external physical signs. In some people, wild physical movements can happen, whereas in others only a flickering of the eyelids is apparent. It is therefore difficult to detect the onset of a fit by measuring physical movement.

Although abnormal brain activity can be observed through sensors on the person's head, having a bath whilst wearing lots of sensors and electrodes on your head is not particularly desirable. My own favourite solution to the problem was the use of a pair of glasses with simple sensors strategically placed. These can measure when the glasses are immersed in water and cause the bath to empty without the need for wires or other sensors. The only intrusion to the person, therefore, is that he or she must wear glasses when in the bath. By the way, the glasses will not steam up, they have no lenses! Meanwhile, a solution suggested by an industrial visitor to the department, obviously with an eye on cost, was to tie the bath plug to the person's toe!

Several current projects on technical aids for people with disabilities are funded by European Community research programmes. One example is the Intelligent Home System, which allows an individual to control all sorts of things around the home from one central position. It uses a technology called LONs (Local Operating Networks) in which each individual item, window, door, television and so on, has its own portion of intelligence, with which it interacts with the central processor and with other items. In this way, the intelligence is spread around the home rather than being at one point. The same method is also being used to control heating around a building, each room having its own allocation of intelligence.

One final project is an intelligent wheelchair, on which we are working as part of a team including the National Technical University of Athens. The aim of this work is to enable a person sitting in a wheelchair simply to request to be taken to another room and for the wheelchair to do the rest. The wheelchair moves around on the basis of a map of the location and must avoid hitting things, including walls and people. It does this in a fashion similar to the electric platform described earlier, although it employs laser light for detecting paths to be taken. This has a longer range than high-frequency sound (ultrasonics).

Project work in the whole area of such technology is exciting, fruitful and richly rewarding. Each case is different and can involve something which is technically demanding. The projects are very much of a cybernetic nature, involving humans and machines cooperating, as one whole. Individually, perhaps both the human and the machine have limited capabilities. Together, however, their capabilities can be considerably enhanced.

The cybernetics department work in this area brought about the link with Sir Jimmy Savile, one aspect of which has already been mentioned. Sir Jim helped to raise £10 million for a new spinal unit at Stoke Mandeville Hospital. As well as our work with disabled people, the link extended to the *Jim'll Fix It* programme, for which we built a 'magic' robot chair[49] which appeared regularly on television for three years.

The chair, which housed amongst other things a robot arm, a full-colour TV, a telephone and a computer, was controlled by Sir Jim through fingertip selection on a touch-sensitive TV screen. It could do all sorts of things, such as make a cup of tea and present badges to participants in the programme. Amusingly, when the chair was first put together, the BBC wanted it, through its computer, to say, 'Jim has fixed it for you' when a badge was handed out. With computer speech now being of high quality this was no problem. The chair sounded just as good as a human. However, the BBC was not happy with this: they wanted the chair to 'sound like a robot'. In the end we recorded the voice of Roger Ordish, the producer, and what actually went out on television was a synthesised, 'robotised' version of his voice. Although computer-controlled, the *Jim'll Fix It* chair could not really be called intelligent in terms of what it actually did on television. However, it has a fairly powerful computer within it and therefore has the potential to do things which could be regarded as intelligent.

Often robot machines have movements which can be related closely to animal or human movements. In consequence, even some relatively low-intelligence operations can bring a message home, and prove a point in a way that a computer simulation never can.

As an example, we connected a pair of cameras to operate in stereo on the end of a robot manipulator arm. The cameras fed an artificial neural network, in the same type of way that humans eyes feed the brain with visual information. With one person as teacher, the neural network was taught the difference between a smiling face and an angry face. As a simple demonstrator, a volunteer, often a potential student, could sit in front of the cameras and pull faces. The neural network simply made a decision between whether the observed face was smiling, angry or neither of these. If smiling, the cameras would move gently towards the individual, and stop. If angry, they would move away from the person. Although fairly simple to set up, the robot could be seen to be interacting with a human and, in this case, to be directly controlled by human facial expressions which it had been taught to recognise.

Recognising patterns of one type or another has been the basis of a number of projects carried out in the department in collaboration with industrial companies. In these cases, an artificial neural network is taught to recognise particular products or articles and to act accordingly. Identical technology can also be used to recognise, for example, different faces. Such a network could thus recognise, and cause a machine to respond to, only men with beards or only blond-haired people.

In many projects within the department a problem is set and the goal is to build a machine that solves that problem. Often there is a practical need at the heart of a project, but in a number of cases it is simply down to pure science, achieving something simply for the sake of achieving it.

One recent project of this type was to get a robot arm to throw and catch a ball successfully, and to do so repetitively. The students who worked on this project called themselves the Omega team, and completed their work well within the one-year time scale allowed, putting in between six and eight hours per week. The finished arm throws and catches a squash ball, or in fact anything else it is given. It has a photodetector in the centre of the hand which detects the presence or absence of an object, by light respectively not reaching or reaching the photodetector. If an object such as a ball is present, the hand throws it up and catches it, and keeps doing so until the object is removed. The next hurdle is to get two arms to cooperate in the ball-throwing exercise in order to achieve a juggling robot.

Other project work has more serious and more directly relevant goals. One such example is Nigel Archer's robot leg, which was not so much aimed at providing a basis for a robot walking machine, but was rather developed in order to study some of the control problems caused by, for example, gravity when walking in different situations. As well as being useful in the study of how robot walkers can be controlled, the leg also proved beneficial in providing information on the type of forces and stresses present in human walking, thereby finding itself a role in the analysis and correction of erroneous gaits in humans.

Another such example is CybHand, which was mentioned briefly in Chapter 4. This four-fingered hand (three fingers and a thumb) has a pulley system which links each finger joint via a string (tendon) to a direct drive motor. Each joint can therefore be separately moved. In actually causing a human finger, or fingers, to go through a desired movement, the joints of the finger are interrelated, with the result that the movements of one joint have an effect on related joints. Overall control of CybHand movements therefore involves, in the first instance, a rather complex set of controlling instructions.

Strain gauges were placed on the finger links and these give an indication of the force being applied to each of the joints when the hand is grasping an object. The strain gauges actually measure a change in the length of each link as the joints move, from which we can obtain an idea of the force on the joints. You can witness this effect yourself by waggling one of your fingers. The skin on your finger links, the main parts of your fingers between the joints, will then become either stretched or relaxed according to the type of movement.

Each of the strain gauge measurements can be fed to an artificial neural network, in such a way that the force applied to hard or soft objects is input directly to the network. The hand is then given an object to hold and forces are applied through the finger joints in order to hold the object firmly. The neural network is taught the relationship between the amount of force applied and the type of object being held. This process is repeated for a whole range of different objects. CybHand's neural network learns just how much force to apply to each object.

Once fully trained – that is, synaptic weightings in the network have been decided – the hand can be given an object, informed what it is, and will then apply the force it decides is appropriate. Two points are of interest, the first of which is that CybHand must be told, in some way, what type of object it is holding. In simple terms this information is input directly to the artificial neural network. A more complex system would perhaps have a vision system which looks at the object being grasped and in turn has a separate neural network linked to it. In this way the vision neural network could learn to recognise the different objects the hand is required to grasp. The vision network weights would be adjusted in such a way that for any subsequent objects the network would decide what object it is and pass this information to the hand force neural network.

At Reading we have not yet linked up a vision artificial neural network to a hand force neural network in this way. CybHand is straightforwardly told which object is being offered and it applies the force it decides is appropriate. The technique, however, of one network to recognise the case and another to carry out the appropriate action is more in line with human or animal neural networks and is the basis of the Seven Dwarf robots which will be described shortly.

A second point is a common issue with all neural networks, whether artificial or real. When a network is taught it is given a range of examples to learn from. This applies to both passive and active learning. Subsequently, once it has been taught, if the network is given an example from within the range, it should have a good idea how to respond even if it has not seen that exact example before. Conversely, if the network is subsequently given an example from outside the range, all it can do is to make a best approximation, which may or may not be suitable.

As an example, assume that a visual neural network is trained simply to recognise different makes of automobile, from pictures. If it is then shown a picture of an automobile it has not seen before, the network can still provide a reasonable guess as to what make it is, from what it knows. If, however, the network is shown a picture of a lorry, it would probably not be able to give a reasonable answer. This principle is applicable to *all* neural networks, whether artificial or real.

CybHand, just like Nigel Archer's leg, is not an autonomous object. They both have electrical power connections and computer processing cables leading away from them. In both cases they are considered to be merely part of a larger entity, simply viewed as separate entities for research purposes. Although exhibiting some intelligence, CybHand's thinking is actually done a cable-length away from the hand. Its thinking is not on-board. Indeed, this is the same in humans. Our hands do not have significant amounts of on-board intellect; that is our brain's role.

Animals, insects and other living creatures are all self-contained beings. True, we need regularly to take in various forms of energy or food. However, for the most part we are completely autonomous. Animals and the vast majority of humans, certain temporary medical cases excepted, do not have obvious wires, tubes and cables coming out of them, connected up to remote buildings in which are sited energy and power supplies and all their thinking and brain power. It is accepted, however, that nowadays humans can remotely log into extra computing power through radio or telephone links, which is going part of the way to our not being completely self-contained.

It is, therefore, of considerable interest to see just how far one can go in terms of creating robot machines which are also completely autonomous and self-contained. Obviously one must allow for the occasional charging of an electric battery, which is probably the best energy source. Occasional links with other robots, humans or remote computers are also acceptable, in order to pass on and receive information, and maybe to be switched on or off!

The programme at Reading has two main strands: walking robot machines and wheeled robot machines. In each case, however, the approach is one of simplicity, a bottom–up approach in which the fully self-contained robots do some things relevant to themselves, rather than a top–down approach which is looking for some overall human goal. Essentially it is looking at robots which are entities in themselves, in that they do not need wires going off to external power supplies or computers, and which have physical capabilities which match well with their intelligence. Our programme in robotics takes things one step at a time and gradually evolves the next robot form from the present one, in terms of solutions to problems, extra features, general improvements and additional functions. At each point in time, therefore, we have one generation of robots, with a related set of capabilities, fully functional and operational, but we also have a further generation at some stage of development.

In terms of walking robots, two are described here, namely Walter and Elma. Walter originated as a student project, and his name originally stood for 'Walking Autonomous Local Tasking Experimental Robot'. However, he is just known as Walter and we never really think about his name meaning anything. Elma, meanwhile, was named after a research assistant in the department.

Walter is a six-legged walking robot, a hexapod, which is based on Local Operating Network (LON) technology. He is completely autonomous, with an on-board power supply and computing power and no external cables whatsoever. LON technology essentially consists of stand-alone neuron-like silicon chips which contain a small amount of processing power and can communicate with other neuron nodes. Walter's design is based closely on the make-up of an insect's nervous system, in that each of his legs is controlled by an individual node. Each of these leg nodes responds to a heartbeat signal sent out from Walter's brain, which is at the back of his head. This heartbeat ensures that leg operations are synchronised. However, it is not critical, in that if it fails for some reason, individual legs continue functioning by each one inferring the operation of the other legs. In this sense, as in others, Walter behaves in a very insect-like way.

Walter's insectlike behaviour was considered further during his cameo role on BBC TV's *Alien Empire* in 1996, a programme which looked extensively at insect life. Walter's structure and operation were compared directly with those of a cockroach. In particular, his individual leg node arrangement was immediately likened to a cockroach's nervous system, to the extent that if one of Walter's legs is removed or malfunctions, he keeps going on his remaining five legs, in a similar fashion to an insect.[50]

The use of LON nodes on Walter's legs allows the control of his walking to be spread about his body in a very efficient way. Essentially each leg takes care of itself, but with regard to what the other legs are doing and what the central brain is telling it. It allows Walter to function with a fairly low amount of necessary computer power for walking. Such an approach is also much more efficient and insectlike in comparison with hexapod robots which are controlled only from one central controller which deals with everything.[51, 52]

Walter's legs are widely spread in order to maintain a low centre of gravity and thus to help stability. Each leg is connected to the body by means of two motors. One allows the leg to swivel backwards or forwards, the other causes the leg to move up or down. Each leg then moves in the same sort of way as an oar of a rowing boat. By synchronising movement of the legs, different gait patterns can be obtained: for example, fast walking, slow walking, turning, shuffling forwards or backwards, left or right.

Each leg node retains a memory of its sequencing needs for each different gait. When Walter desires to move in a particular direction and with a specific gait, this is accomplished by means of a simple command from his brain to the legs. The movements of the legs are then coordinated by the legs themselves, and Walter can content himself with thinking of higher things. This mode of operation allows for a high degree of robustness and reliability.

As well as selecting a particular gait and causing an instruction about this to be passed on to his legs, Walter's brain also carries out a number of behavioural responses. In the first instance, he has an infrared receiver, which is actually situated where his bottom should be. This is used to send him information. Although potentially very powerful, we have actually used this link mainly for research and demonstration purposes, switching Walter on and off (in the same way and by the same means that a TV is switched on and off remotely) and remotely selecting gait patterns and directions of walk.

Walter also has ultrasonic transmitters and receivers on his face, to the right and left, rather like an insect's eyes, but dealing with sound rather than light signals. In this respect his senses are more like those of a bat than an insect, in that, as mentioned in Chapter 4, bats manoeuvre safely by detecting objects or other creatures through emitting a high-frequency sound signal. If an echo is received the bat knows something is nearby and has a good idea of its distance. Walter sends out ultrasonic signals from both the right- and left-hand sides of his face. If an echo is returned he knows that something is there, to his front left or front right or both, and he has a good idea of its distance.

Once started off, by an infrared starting pulse sent to his bottom, Walter marches forward in a selected gait, again chosen through the infrared start-up, until he encounters something in front of him. If he approaches an object, whether moving or stationary, he can avoid it in a number of ways. Apart from not hitting things, he can be set a goal such as to remain a minimum distance from objects, which he can accommodate immediately. Essentially, when he encounters an object, Walter must take a decision as to what action his legs should carry out: to go backwards, continue forwards, spin to the left and so on.

Rodney Brooks,[53] a Professor at the Massachusetts Institute of Technology's Artificial Intelligence Laboratory, referred to the different basic behavioural traits of an insect as levels or layers of operation. They are, with the lowest level first:

1. Avoid contact with objects.
2. Wander around, avoiding obstacles.
3. 'Explore' the world by setting distance as the main goal of the behavioural strategy.
4. Build a map of the environment for use with path planning.
5. Notice changes in the 'static' environment.
6. Reason about the world in terms of objects and perform tasks related to the objects.
7. Formulate and execute plans that involve changing the state of the world in a desirable way.
8. Reason about the behaviour of objects in the world and modify plans accordingly.

In this sense Walter merely has the capabilities to accomplish about three levels of behaviour and is not yet able to achieve level 4. It has been found at Reading that dealing well with the first three levels, completely autonomously, still provides a considerable challenge. However, Walter is also able to communicate, and this, I feel, is important, even though such a capability does not fit immediately into the basic Brooksian hierarchy.

Considerable improvements have been made to what was learned in the design, construction and operation of Walter, and Elma has emerged as a result. Elma has many features in common with Walter, including her entire nervous system arrangement in terms of a LON-mode implementation. Her potential walking gait, ultrasonic detectors and infrared links are also almost the same. Elma is superior, however, in a number of ways, and she is consequently a new generation, a step on from Walter.

Firstly, Elma's body is a sleeker yet sturdier design, resulting in much better stability, particularly when the fastest walking gaits are required. When Walter moves at high speed his stability deteriorates and he almost falls over. Elma has no such problems.

Elma also has contact sensors on each of her six feet, so that she can feel when she is actually touching the ground. As a consequence she can rearrange her stance to retain her balance even when her legs are positioned awkwardly due to uneven flooring, objects in the way or when stairs need to be climbed. Each leg is able to collect information on whether or not its own foot is actually touching the ground and can pass this information back to Elma's central processing brain, which contains around 100 neurons, and which organises a body-balancing operation.

Elma has been given a radio link which enables her to communicate directly with a computer. By this link she can send and receive information on her environment, which can be used at the computer end, if required, to build up a three-dimensional mapped picture. Instructions can then be passed back to Elma either to reach a certain place on the map in a reasonably quick time or to avoid a particular area. In due course, however, it would be preferable to have such planning capabilities on board, rather than a radio link away, but this will probably be part of the next generation of walking robot.

One final capability which Elma has is that of learning, which is again outside the Brooksian hierarchy. In this way she can potentially learn how to move around without hitting things and even learn how to walk. However, our research into such actions is in the relatively early stages as far as Elma is concerned. This is discussed further in Chapter 11.

Learning is, however, a feature of the wheeled robots which have been developed in the department, although it was not part of the first generation's capabilities. The original robots were given the nickname of the 'Seven Dwarfs', probably due to the fact that they are small and cute and there were originally seven of them. With different generations and special designs there are now around 20 robots, but we still affectionately refer to them as the Seven Dwarfs. Although neither Snow White nor the Wicked Queen has actually appeared in their entirety, both have, as will be described shortly, been duly represented in the lives of the dwarfs. So let us have a look at the first generation of the dwarfs.

The robots move around in an environment such as the floor of a room or building, by means of two motor-driven wheels which can go forwards or backwards at varying speeds. These wheels are positioned at the rear of each robot, to the left and right. At the front, each has a simple caster wheel with no driving or steering mechanism. This wheel is simply there to run freely and to keep each robot steady and upright. Due to a very low centre of gravity, we have no problems whatsoever with the robots' stability.

It is worth stressing that all of the wheeled Seven Dwarf robots are autonomous robotic organisms in their own right. They are individually self-contained, with their own rechargeable on-board battery supplying all the power required to drive the wheels, to service the robots' electronics and brain and to power all the sensing devices. Under normal, active behaviour, a fully charged battery will provide enough power for about six hours. It has taken some time for us to get to this stage, however, as initially we could only achieve about 15 to 20 minutes.

The programme of work on the Seven Dwarfs started in about 1991 with some initial designs, usually built into prototypes purely to see what would happen. Even before the first *official* generation of Seven Dwarfs there were, therefore, some trial versions from which we learned. The robots today have fairly powerful motors and move around quite quickly. In the pre-Seven Dwarf era, however, even more powerful motors were used, coupled with open gear boxes. These first robots would hurtle around the laboratory at breakneck speed, crashing into walls and doors. They were, therefore, designed with metal bumpers at the front and with a sturdy frame. Another early problem to be overcome was that of stopping a robot once we had started it. Catching a rapid transit autonomous robot with an open gear box, in full flight, was quite dangerous, with a serious chance of injury. More often than not we simply had to wait until the robot's battery had run flat and then reclaim it.

So the first generation proper of wheeled robots had slightly less powerful motors, with enclosed gear boxes and rugged body frames. The basic design, just as in the case of Walter and Elma, was centred on the use of ultrasonic eyes, on the front right and front left of each robot's face. As with Walter, each robot can then sense the presence or absence of an object to its front right or front left. If something is there it can work out its distance by measuring the length of time taken for the sound signal to be returned.

The control problem with the robots can be conveniently separated into three portions: sensing, characteristic behaviour and motor control. On the sensing side, a decision needs to be taken by a robot, along the lines of 'Is there an object to my front left or front right or both, and if so, how far away is it?' In the first instance this relationship could be, and indeed was, built up on a simple programmed array, in such a way that signals from the ultrasonic sensors were converted by the array into a direct answer to the robot's question.

The second part, the characteristic behaviour, was based on computer memory, which can be directly erased and programmed. This part takes sensor information and converts it into commands for the wheel motors. For example, if an object is a medium distance away from the front right eye then drive the right wheel forwards at medium speed and the left wheel forwards at slow speed.

The final part was then simply to take the commands from the characteristic behaviour memory and convert them into voltage signals for the wheel motors by means of another programmed array. In this way the wheel motors can independently be driven forwards or backwards at a desired speed.

With the programmed arrays for both sensor and motor control fixed, it is, therefore, a straightforward case of developing a behavioural response for a robot, in software. This is directly downloaded into the erasable memory. One may then see what happens when the robot behaves in the way you have programmed it. This facility is used by us at school fairs and exhibitions, so that schoolchildren are given the opportunity to participate and actually program a robot brain and watch it behave in the way they decided. The children get very attached to their own 'offspring', particularly if the robot does not do what they thought they had programmed it to do.

Although different behaviours of the first generation of Seven Dwarfs are considered in the next chapter, one straightforward way to programme one of the robots is as follows:

1. If no object is in front, then drive both wheels forward at high speed.
2. If an object is in front and to the right, then turn left (away from the object).
3. If an object is in front and to the left, then turn right (away from the object).
4. If an object or objects are detected in front and both to the left and to the right, then see which case of 2 or 3 is strongest and follow that rule. Otherwise spin round 180°, or as an alternative, reverse and spin 180°.
5. On completion of any of cases 2 to 4, return to case 1.

When operating in this mode, a robot simply moves around, generally forwards, avoiding any obstacles it comes across or which are put in its way. The extent of motor driving and the sharpness of the turn in cases 2, 3 and 4 is dependent on how close an object is. The overall objective of the robot is to move at such a speed and with such a response that it never hits anything. Other extremely interesting behaviours are also possible, however. Some of these are considered in the chapter which follows.

One problem with the original design was that each robot's on/off switch was positioned next to its head, which had initially seemed a good idea. Unfortunately, on approaching a robot to turn it off once it was running, it would detect a hand coming near it and move away from the person trying to switch it off. So, once again, we simply had to wait until the robot's battery ran out. In the second generation of the Seven Dwarfs the on/off switch was positioned at the back, and in the third generation, underneath.

Another problem to be overcome in the first few months was that we kept losing robots once they were let loose in the laboratory. A robot would simply head off in one direction, avoiding things, and if we did not watch where it was going it could disappear under a table where it might well remain if its battery went low at that time. On a number of instances a robot went missing only to be discovered a few weeks later in a remote corner of the laboratory. Happy, one of the brighter Seven Dwarfs, was even discovered in the computer science department, which is adjacent to our own. He had travelled through several fire doors and along a number of corridors before being waylaid.

In their first year of studies in cybernetics, students get a chance to try all sorts of programs in the robots, as part of their practical work, and to see how the robots respond. As you might expect, many strange behaviours have emerged. In order to stop the robots heading off, a corral has been built for them, and most practical work and demonstrations are contained within it. A number of special features and technical designs have been included in the first generation of the dwarfs, and these are discussed in Chapter 10. However, a distinct step forward was first taken by Bashful when he was given the capability of learning. He was provided with a new brain, an artificial neural network, replacing the erasable memory and programmed arrays.

When initially switched on, Bashful has no in-built behaviour pattern, but has to move around in the corral, learning how to move his wheels. His basic, programmed instincts are to keep moving and to avoid obstacles, but he has to learn how to achieve these through active learning, by trial and error. Bashful's aim is, therefore, to learn general behavioural responses which allow him to perform well in his environment. There are many parallels between Bashful's learning and learning in children, as will be discussed later.

The third and most recent generation of the Seven Dwarfs incorporates communication between each other. Each robot has a ring of infrared transmitters on top of its head, with infrared receivers between. Each robot transmits its own unique signal frequency, a carrier frequency, on which are superimposed pulses of information. In this way each robot is readily aware of the other robots, because of the signals they are sending out. The behaviours already obtained from these robots have been eye-opening and in many cases very unexpected, particularly where parallels with human or animal behaviour can be observed.

The purpose of this chapter has been to introduce some of the robot machines developed in Reading University's cybernetics department to provide a basis for studying learning behaviours and investigating machine intelligence. Each of the projects mentioned adds in its own way to the bank of knowledge obtained. The emphasis, however, is most definitely placed on actually implementing, on applying, the intelligent techniques discovered, thereby finding out far more than was imagined possible. The proof of the pudding is not just in the eating, when the pudding turns out to contain unexpected fruits.

The Seven Dwarf robots considered so far, all built in the department, are very good bases to use for the development of learning strategies and the study of different behaviours. Though they are harmless, fun and friendly – just as all the researchers in the department are, including myself! – robots do not necessarily need to appear like this.

We have also built in the department a 5 ft high autonomous version of one of the Seven Dwarfs. This particular robot is based on a science-fiction robot, a Dalek, and does not itself appear as immediately friendly. Visitors' reactions to it are not at all the same as they are to the Seven Dwarfs. The technology is the same, it could learn in the same way as Bashful, and it can avoid hitting things, yet because it is much bigger and quite powerful it generates mixed feelings. This robot too will be considered further in the next couple of chapters, rather in terms of what it could potentially do, given today's technology, than what it can do now.

Chapter Ten

Our Robots Today

In cybernetics at Reading we receive very good funding to support our research activities. With finance obtained from UK government research councils, UK government departments such as the Department of Trade and Industry, European Community research programmes and international industry, we are able to ensure a good-quality research structure including the latest technology and equipment. To serve as examples, in recent years industrial support has come from BP, British Gas, National Grid, Unilever, British Telecom, Sony, the Bank of England, SmithKline Beecham, and Satchwell Control Systems Ltd (GEC). SUN Microsystems (USA), the world's leading computer workstation manufacturer, have perhaps been our biggest supporter.

Much of the work carried out involves either the design of new theory, with supporting simulations, or applying the intelligent methods we have originated to real industrial problems, for the benefit of the companies concerned. In each project, however, the sponsor, whether government or industrial, has considerable interest in the direction of the research and therefore makes demands in terms of what research is carried out and what must ultimately be achieved. In other words, we have to toe the line. There is some flexibility, but not much.

Our programme concerning intelligent robot machines is run differently. In this case, deliberately, we in the department are in control. It is we ourselves who decide which research direction we are going to take, how quickly we achieve our targets and whether the direction of research should change because we see the possibility of a leap forward. The research work is consequently pioneering, exciting, enormous fun and gives the most gigantic kick when a major breakthrough occurs.

Resources are channelled in the department to support the work, although, as is described in this chapter, the research does not actually require large amounts of finance. Our key asset is not expensive supporting equipment, or an open chequebook from a mysterious source, but a wonderful group of clever researchers who are all fired up on the way the project work is going. Indeed, high praise must go to Dave Keating, a lecturer in the department, who heads up the design and building of the Seven Dwarf robots, for the way he has helped the programme ever onwards.

As we are not directed by defence concerns, either government or industrial, we are not in any way building machines of destruction or war. Rather we can investigate what is possible in terms of intelligent robot machines and realistically look at what could be possible in the future. In this way we can keep a check on what is going on now and what could be going on in terms of development in the field. We therefore see ourselves as the 'good guys', or perhaps more appropriately 'white knights',* reporting openly on what is and is not possible in the field of intelligent machines.

It is quite possible, when researching into intelligent machines, to follow a set of five instructions as follows:

1. Buy an expensive, extremely powerful computer.
2. Buy an expensive, extremely powerful video camera.
3. Connect the two things up together with some expensive interfacing.
4. Try to get the computer to do something with the picture coming through from the video camera; for example, understand the differences between a tree and a wooden post.
5. Use the computer output – for example, 'Yes, it is a tree' – to do something, such as steer a vehicle.

My own view is that such an approach is invariably excessive in

* Whiteknights is the name of the Reading University campus.

terms of what is actually required to tackle the problem in hand, and often tries to make a machine operate, unnecessarily, with human values and human concerns. As an example, if we are trying to get a computer to drive a vehicle around an object, the computer probably does not need to know, or understand, the difference between a tree and a post. It may simply need to know that something is in the way, but not what it actually is. For this a video camera may or may not be the best approach. Visual images, in their primary form, are relatively complex things which can take a machine some time to process. If it is required for a vehicle to travel at 60 miles per hour, it is no use if an on-board intelligent system takes 25 minutes to decide whether the object the vehicle is approaching is a tree or a post. On such a vehicle the best result we could hope for is a response 24 minutes, 59 seconds after a crash that says: '25 minutes ago we crashed into a tree.' If the computer is still operational after the crash, that is.

Would it not be much better, in the case of a computer-driven vehicle, for the computer to be able to decide quickly, 'There is something in front of the vehicle so we must steer around it, or slow down'? In this case, if a visual image is used, it could simply be employed to indicate that an object is in front of the vehicle and, if appropriate, how wide that object is.

In the biological world, insects and animals, including humans, have evolved in such a way that in each case their physical capabilities and potential go hand in hand with the amount of brain power provided, in terms of both size and complexity. So a slug's brain, albeit fairly small, is appropriate for a slug. Similarly, a dog's brain is appropriate for a dog and a human's brain is appropriate for a human. In a human, for example, it is estimated that over two thirds of the information entering our brain is concerned solely with vision.[54] In each case, mental and physical attributes have evolved to a balanced form. Thinking about a slug's brain in a human's body, or conversely a human brain in a slug's body, indicates how problematic an imbalance would be.

Our own approach to intelligent robotic machines has kept this balance clearly in mind. We have attempted to match the available power and capabilities of each robot brain with the physical role it has to play and hence with the senses necessary for that. Where only one or two simple tasks are to be carried out, only a fairly simple and relatively small brain is required.

One aspect of this, which is of considerable interest to us at Reading, is collective intelligence. Consider, for example, a honey bee. Each bee has a brain containing thousands of neurons. As an individual, a bee can accomplish a fair amount, not only flying, feeding and self-preservation, but also remembering where certain flowers are and when to fly to these flowers dependent on the time of day. Even more significantly, the bee can communicate with other bees and tell them where particular flowers are. As a collective, bees can also successfully build and operate a hive.

Through communication a group of bees is cleverer than an individual bee. But ten million bees together are not, one feels, as intelligent as a human, even though their collected total of neurons may be similar. Clearly communication adds another dimension, but because of the strict partitioning of neurons into individual bee heads, a complete, complex connection of all of the neurons in ten million bees is not obtained. The concept of universities is similar, in that putting a number of humans together, for discourse and shared learning, has a much more positive effect than if all members stayed at home and ignored each other.

As humans we have, throughout history, compared animals with ourselves, finding it quite amusing when a chimpanzee does something in a humanlike way, such as clapping its hands. We have attempted to get animals to behave as we direct, not only Pavlov's dogs but also getting bears to dance or lions to stand on podia or birds to make human noises. When they do so we ascribe a certain amount of intelligence to them. A bird which can say 'Reading University is best' is more intelligent than a bird which cannot talk at all.

We ascribe human values and human names to animals and assume that they like us or have other humanlike feelings. When considering machines which move around and respond to the outside world we tend to do the same. As an example, maybe you have sat in your car, which will not start, and shouted at it, 'Come on, you ****, start!' or, 'Please, please start, I'll be late if you don't.' Does the car understand you?

We should be careful when we attribute human values or human understanding to non-humans, whether animals or machines. Nevertheless we humans do it, and I am no exception. I am human, after all. What is described in the remainder of this chapter, the results from our robotics work, is looked at very much from a human value point of view, not only in terms of the robot machines themselves but also in terms of how the behaviour of the machines compares with human or animal behaviour.

In the first instance, let us consider the first generation of Seven Dwarf robots. These are, essentially, only programmed devices. In human terms they only have basic instincts and cannot learn or change their characteristics in any way other than by a malfunction or by their batteries running low. The only sensory information they obtain is the presence or absence of an object to the front left and/or front right.

One basic program form is that of self-preservation. If an object is in the front, to the left, then the robot's wheels are operated to make it turn right. The converse applies when there is an object in front, to the right. The robot, therefore, is not only programmed to move forward, but will also turn away and avoid objects put in its path. This same instinct can be found in any animal but is most clearly shown in an insect. Next time you try to swat a fly, watch what it does.

But these robots can be programmed in different ways. One alternative is for them to move forwards, but to stop when an object is encountered. If the object then moves away from it, the robot will move forward again, tracking the object, but getting no closer than a preset distance. An interesting behaviour is then obtained when the Seven Dwarfs robots are placed one behind each other. If a moving object is put in front of the leading robot, that robot will simply follow the object around, with a string of other robots trundling along behind, each one following the robot directly in front of it. Because of the way the ultrasonics are placed on either side of each robot's head, they tend to move slightly from side to side as they track the one directly ahead.

When watching the Seven Dwarfs performing in this way, it is tempting to liken their action to that of ducklings, waddling along one after the other behind their mother. We feel that the ducklings are being quite clever when they do this, and yet for the robot machine it involves nothing more than a few lines of program, a simple basic instinct. It is said that ducklings will follow the first thing they see when they are born. In the case of the robots, however, in this mode they follow absolutely anything of a reasonable size. They will even follow a wall, which, as you might guess, is very unlikely to travel far!

Another extension to the program causes the robots to remain stationary, except when an object is a reasonable distance, approximately a foot, in front of it. Then, when the object moves, the robot will remain the same distance away from the object. If the object moves forwards, the robot will also move forwards, whereas if the object moves closer, the robot will back off to a 'safe' distance. The 'object' in this case can easily be a person walking. The robot follows along behind the person, remaining a short distance away. If the person moves towards the robot, then the robot will back off, giving the appearance of being 'unsure'.

This simple mode of programmed robot operation appears very much like the behaviour of a dog, in particular a puppy, following along at the heel of a person. In the case of a puppy, such behaviour is often regarded as indicating how clever it is. Yet with the robot it is once again merely a line or two of code, a basic instinct. As I shall describe shortly, such a behaviour could just as easily be learned by the robot, as long as we are able to reward it appropriately. Clearly it becomes very difficult to say in the one case, for a dog, that such behaviour is clever or intelligent without exactly the same being true for the robot. We are witnessing behaviour and from this are making a judgement on the intelligence of the thing, machine or dog, doing it. This problem is the same as is raised when trying to compare the intelligence of humans and machines.

Everything that the first generation of the Seven Dwarfs can do, so too can Walter, the hexapod robot, although Walter of course walks. So Walter can walk along behind a person, like a pet cockroach rather than a pet dog. However, he is somewhat slower than either a cockroach or one of the Seven Dwarfs, so the person leading him needs to walk fairly slowly. When he is placed in the corral with the Seven Dwarfs they all avoid hitting each other as they move around, although as Walter moves relatively slowly he tends to saunter from one side of the corral to the other, with the Seven Dwarfs scurrying around him.

The Seven Dwarfs simply move around on their wheels, which they can independently drive forwards or backwards at different speeds in response to external circumstances. Walter, meanwhile, has different gaits, different ways of walking. It is a straightforward exercise to programme Walter so that he can walk in different ways depending on what is happening around him. Not only can he move sideways and/or backwards in order to avoid or move around things, he can also alter the sequence of leg placements as external conditions change.

While still considering the first generation of Seven Dwarfs, a number of additional features were given to some of them. In fact these features can also exist in the later generations, if required, though for demonstration purposes they are not usually retained.

The first type of extra response is that of predator or prey. As well as its normal faculties, a robot is fitted with a receiver to enable it to detect infrared signals. The robot is programmed to respond to two different infrared signals, at different frequencies. It responds to one, the predator signal, by moving away as quickly as it can. When it gets far enough the signal becomes weak and the robot can return simply to moving around, avoiding obstacles. However, as soon as it gets near enough to the predator again the robot picks up its 'scent' and once more heads away until the 'safe' distance is reached.

With this action a robot is exhibiting different levels of behaviour. It has its normal, basic level of simply moving around avoiding things. However, if a predator comes near, the robot makes it a priority to get away as quickly as possible. In terms of the Seven Dwarfs this behaviour can be called the 'Wicked Queen' effect. I should add that although I have described this behaviour as being programmed, it could just as easily be a behaviour that is learned by the robot. As long as the robot is penalised in some way for getting too near to the predator then it can learn, in future, not to get too near again. Such a method of learning will be described shortly.

In Chapter 11 I describe how one robot can teach another. It is apparent, therefore, that the self-preservation characteristic in the presence of a predator is something that one robot could learn from another robot. Another approach is to assume that we build ten robots, five of which hurry away when a predator approaches, whilst the other five do not. If the predator really is nasty, then the five that do not hurry away will not last long. Any future robot designs will, we assume, be based on the five remaining robots, all of which can escape from predators. This is essentially genetic engineering, ensuring that all future robots are programmed with the basic instinct of escaping from predators. A similar discussion was, in fact, held earlier in Chapter 8 with respect to humans, the predators on that occasion being tigers.

One could, of course, argue that a predator would not necess-arily emit an infrared signal of a specific frequency. Most likely it would not! The robots are, however, simply being used to prove a point. The robot, or indeed a human, must have some way of identifying a predator and of detecting when one is approaching. A human would probably use a visual picture. As in the example considered previously, if we see a tiger coming towards us our instinct is to run away. With the robot, the infrared signal is merely being used to show what is possible, based on the robot's senses; it only has ultrasonic and infrared sensors. We could easily put some simple optical sensor on a robot so that it responds to a sudden change in light intensity as indicating a predator. Not only insects respond to sudden light change; even humans react to it with wariness.

The other infrared signal, of different frequency, is used to indicate prey. In this case the robot moves towards the source which emits the signal, and if the source moves, the robot chases after it. In Seven Dwarfs terms this could be regarded as the 'Snow White' effect, in that the dwarfs would group around Snow White if she is present.

The same discussion could be held in terms of the robot's prey as was held for its predator. In normal life the robot busies itself and only responds to the prey when it comes into range. In this way it could be considered that the robot spends its life moving around, and when it senses prey nearby, by detecting its infrared signal, it chases after it.

But what does a Seven Dwarf robot do when it actually locates its prey? The prey, as described here, is merely a piece of electronic circuitry which is emitting an infrared signal. The robot cannot actually eat the prey or get energy from it, can it? The answer is yes, it can certainly get energy from it, as follows.

The Seven Dwarfs make use of rechargeable batteries. It is not too difficult for a robot to check on the status of its own battery; the best method is simply to measure the current drawn from the battery over a period of time to see how the electrical charge in the battery has dropped. Indeed, this is exactly what has been done. Further, the robot is fitted with two bumpers at the front, which can make an electrical contact.

What happens is that a robot moves around in normal mode until it eventually detects that the state of its battery is below an acceptable level. The first one given this ability, incidentally, was called Doc. At this time, and not before, his priority is to get his battery charged. Positioned elsewhere is a station at which the robot can charge itself, and at this station is a beacon which emits an infrared signal, rather like the signal indicating prey. Once the robot has decided he needs to charge his battery, he starts to look for the nearest beacon. As soon as he locates it, the robot heads towards the beacon.

Once Doc arrives at the recharging station he circles around it until he gets to the point at which the signal is strongest. At this time he moves directly towards the beacon, thereby docking (see where his name came from!) with the recharging station. The robot's battery is then charged through the electrical contacts on its front bumpers. Once the battery is fully charged, or a specified charging time has elapsed, Doc reverses from the charging station and resumes his normal activities. This mode of maintaining energy levels is perhaps one of the most difficult, in that the robot must be initially aware that its own battery is fairly low and needs recharging, and then it must locate a charging station, which could be anywhere.

A much simpler approach would be for the robot to appear at a charging station in a known, fixed location, every time a preset time period has elapsed; for example, once every three hours. However, at Reading we have taken the more difficult route and have robots which recharge themselves when *they* decide that their batteries have gone low.

As an aside, it is easy to look ahead from this to completely self-sufficient robot machines. All that is required are some electrical charging stations placed at various locations, rather like McDonald's for robots. But let us get back to what our robots can do at present. Let us look at the second generation of the Seven Dwarfs, the robots that can learn.

As described earlier, a first-generation dwarf, in its simplest form, is programmed so that if an object appears on its front right or front left, its two wheels will respond in an appropriate fashion. In this way the robot avoids objects but keeps moving. These two goals, to avoid objects and to keep moving forwards, were carried over into the second generation of Seven Dwarfs. However, that programmed response has been replaced by a learned strategy, which aims to meet the two goals. A direct comparison can be made with a human child, which has a goal to walk without falling over. A child has, after some time, all the necessary physical hardware to do the job, but must learn mentally how to walk through active trial and error. It is interesting to consider why humans have evolved in such a way that we must learn to walk, whereas in other mammals walking either appears to be a basic instinct or is learned very quickly.

A dwarf has the necessary hardware, in terms of wheels to rotate, but we have arranged things so that they must learn how to move around. Bashful was the first of the Seven Dwarfs to have his purely programmed response replaced by a new brain, the learning aspects of which operate as follows:

The overall problem is split into two parts, the first being an indication of what situation the robot is in and the second being a decision as to what to do about it. I shall describe the situation indication section first.

What a robot is facing can be split up into a large number of different scenarios. However, to keep things reasonably simple, Bashful was given just five different situations to select from at first. These were:

Situation 1:   No obstacle detected
Situation 2:   Obstacle near to the left eye
Situation 3:   Obstacle near to the right eye
Situation 4:   Obstacle distant from the left eye
Situation 5:   Obstacle distant from the right eye

From his ultrasonic 'eyes' Bashful receives signals indicating whether or not anything is in front of him. The signals are, however, not clear 'yes' or 'no' indications but rather are typical of everyday life. They only give a rough guide, from which a decision must be made as to which one of the five situations outlined is most likely to be the actual case at any time. An artificial neural network containing only 40 neurons, based on a Hopfield network,*[55] organises itself, that is to say it arranges its synapse weightings, in order to indicate which situation exists at any time, given the signals coming in from the ultrasonic eyes. This self-organisation occurs as the robot moves around, obtaining values from its environment.

The operation of this type of artificial neural network can be likened to the arrangement of a map. When certain situations occur, specific groups of neurons will be more active than others. This means that each particular situation is reflected by activity in a corresponding region of the map.

The second part of the problem is for Bashful to decide what he should do, given that he is in a particular situation. To start with, Bashful was allowed nine possible different actions corresponding to each of his two wheels being able to move

---

* A Hopfield network is essentially an array of neurons in which the signal output from each neuron is fed back to act as an input signal as well. It is one particular type of neural network, that has been found to operate well when information needs to be classified into one type or another. It is therefore extremely useful for classifying Bashful's situation.

forwards or backwards or to stop. For example, one action is right wheel forwards and left wheel stop, and another is right wheel backwards and left wheel backwards.

Each of Bashful's possible actions has a weighting, a simple number, associated with it. However, a separate group of weightings is given for each of the five situations. For a particular situation, at a particular time, the action with the highest weighting can be carried out. If the action is successful, in that Bashful moves away from any obstacles, then the weighting of that action is increased and the other weightings are decreased. If it is an unsuccessful action, in that Bashful moves nearer to an obstacle and maybe even hits it, then the weighting of that action is decreased and the other weightings are increased. In this way Bashful can move around in his environment and learn, firstly, what situations exist according to his ultrasonic eyes, and, secondly, what actions to take for the different situations.

In fact, for a particular situation, rather than simply carrying out the action with the highest weighting, what actually occurs is that a biased roulette wheel procedure is used. In this way all the different actions are likely to be tried at some time, although the action with the highest weighting is that which is most likely. If the highest-weighted action is tried and is successful then it is even more likely to be tried next time, whereas if one of the low-weighted actions is tried and is unsuccessful then it is less likely to be tried next time. Hence, as learning proceeds, it becomes more and more likely that particular successful actions will be carried out for each of the five situations.

The hope is that for situation 1 Bashful will eventually learn that the best action is right wheel forward and left wheel forward, thus moving him forwards when no obstacles are detected. In this case he is actually encouraged to move forwards by a direct reward; that is, within situation 1 it is inherently indicated as a good strategy when both wheels move forwards. For the other four situations there is a range of possible solutions which will cause Bashful to avoid hitting obstacles, and one can never be sure which actions Bashful will actually learn. Indeed, for a particular situation Bashful may end up by switching between two or three strategies.

When Bashful is first placed into an environment, usually the corral, the weightings on possible actions and his artificial neural network are set up pseudo-randomly. That is, the weightings are different from each other, and whilst they appear to be random, have actually been artificially set up and are hence called 'pseudo random' as explained in Chapter 6. So when he is switched on, any action is possible, though he quickly learns some responses. Forward movement is usually one of the first. However, the strategies he learns when objects appear to the left or right are directly dependent on the successful attempts he makes early on. If, when first approaching a wall, he decides it is nearer his right eye, then he may learn a strategy of stopping his left wheel and spinning around by moving his right wheel forwards. He may, however, learn to do completely the opposite in this situation, but this is less likely.

Essentially, once Bashful is switched on, he will learn a number of actions which he relates to certain situations, but due to several factors, he will be a different Bashful each time he learns because he will exhibit different behaviours. How he is positioned at the start of his life and what he encounters early on are important considerations. In particular, what strategy he first tries out successfully when faced with an environmental feature will have a considerable effect on his later behaviour. For instance, if he moves towards a tight corner and successfully gets out of it by reversing and spinning to his left then he will probably keep this characteristic for the rest of his life, provided it continues to be successful. Bashful learns by trial and error. Only by hitting walls, probably on several occasions, can he learn that it is not good to do so.

The way Bashful's pseudo-random neural network weightings are originally set up is also critical, as this will cause him initially to try out certain actions in preference to others. This governs his initial behaviour, in terms of what actions he tries, the strength and connection with which he tries them and how long he tries unsuccessful strategies before trying something else. The way the environment is set up is also vitally important. Bashful needs freedom to move around and learn, plenty of space but also a rich environment in terms of the shape of the corral wall.

What Bashful is learning is not a map of the corral and the

obstacles within it, but simply a number of general behaviours which allow him to move around in an environment without hitting things. To achieve this he learns to make associations between the signals received by his ultrasonic eyes and whatever successful actions he carries out. This basic form of learning can then be carried over to other activities. For example, when Bashful also makes use of an infrared receiver, he can learn where to go to charge his battery, by means of the characteristic frequency of the infrared signal sent out by the charging station.

In 1994 Bashful appeared on the BBC's *Blue Peter*. Unfortunately one of the presenters tried to 'mother' him far too much, by crowding him and not allowing him to move far before putting her hand in the way to stop him. As a result Bashful was extremely confused and was hesitant about moving in any direction. He was, however, helped by another presenter, who opened things up for him and at least gave him a reasonable chance to learn.

Bashful learns different strategies each time he is given a life. At times he has been known to go into little spins, to spend some time simply moving forwards and backwards from a point on the corral wall, or on occasion just to stop altogether. This latter behaviour usually happens when he has been deliberately confused by having been made to collide with an object whatever he does. As a result he appears just to give up on life. His basic goal of moving forwards is counterbalanced by his learning that if he does not move then he will certainly avoid collisions.

Usually the corral does not contain any moving things when Bashful is learning to move around, other than Bashful himself. This is because Bashful could gain the impression that a good way to avoid colliding with an object is to move towards it quickly, since when he reacts this way to some moving objects, such as another of the Seven Dwarfs, they might move away from him. Unfortunately, if he moves quickly towards the corral wall, he will probably not avoid it! Encountering moving objects too early in his life could therefore leave Bashful with a slight hesitancy, because he has received conflicting reports regarding what is, or what is not a good strategy.

There are many parallels here with human learning: the importance of a good environment, without conflicting influences; the importance of encountering a particular situation at a suitable time; the strong influence of initial behaviour, arising from the individual's gene program; the strength of particular basic instincts and the talent to do certain things well right from an early age. It is easy to see links with humans in terms of positive behaviour, but we can also see relationships with hyperactivity, children's traumas and even criminal behaviour.

The fact that Bashful has particular talents can be exhibited in a number of ways. Firstly, a strong bias can be placed on the way he tries certain actions, by adjusting network weightings so that a little success gives him a strong preference for particular strategies. On the other hand Bashful's network weightings could simply be biased at start-up, so that he is more likely to behave in a certain way early on. In this respect it could be considered that if Bashful's original selection of neural network weightings is 'better', then he is subsequently more likely to perform 'better' in his related behaviour. Overall, therefore, Bashful's performance depends on his initial gene program, the environment in which he can learn, and also his physical aptitude, in that if he has some physical problem this can have a considerable effect on his behaviour.

The importance of the environment on Bashful and which actions he tries at particular times directly affects the way in which he develops. On one occasion he can be started off, learn quickly and appear extremely confident as he moves around, rarely running into problems. Yet shortly afterwards he can be restarted as effectively a new Bashful, having to learn afresh, getting into some difficult situations and never really performing well. This indicates that if we humans could simply go back to the day we were born and live our lives again, we would almost certainly be quite different due to our experiences. This would be even more so if everyone else was starting afresh, because our interactions with them would be completely different. Unfortunately, as yet we cannot turn our own clocks back, but we can cause Bashful to be a baby again simply by flicking a switch.

This behaviour has been demonstrated to audiences around the world including, in February 1998, the World Economic Forum at Davos, Switzerland, attended by among others Helmut Kohl, Bill Gates and Hillary Clinton. At these events a particular experiment is carried out. Firstly, Bashful is demonstrated learning in the normal way, for this he is placed in an uncluttered corral environment. He is then removed from the corral and several programmed robots are placed in it, Bashful is then returned, ready to learn again, this time his life being made far more difficult by the conflicting information received from the other robots moving around him. This type of 'childhood' environment can produce one of two different outcomes. In some cases Bashful becomes a self-assured, determined 'adult' robot, who makes clear and decisive selections. Whereas in others he tends to be extremely hesitant and far from self-assured, with his decision making being less than clear, the machine equivalent of a nervous wreck. It becomes clear from this experimentation that a poor start in life does not automatically mean a poor 'adulthood', we can never be sure which way the robot is going to be.

When he is first turned on, Bashful has some essential goals in life, but he also has a number of basic instincts. The way he starts his life, what he does first and how he does it, is indeed programmed and yet also depends on the environment he first encounters. The way in which he learns is also instinctive, although it is possible for him to acquire different or better ways of learning if his learning rates are adjusted in response to some stimulus.

The rate of learning is an interesting factor. It has been found preferable for a high rate of learning to be applied in early life but for this to get less and less as time goes on. In fact, this tends to occur anyway when certain strategies prove to be successful time and time again. However, weightings of other actions are only reduced proportionally and never become zero. In this way it is always possible, although very unlikely, that Bashful could try a different strategy in a particular situation. It would be possible to *freeze* the weightings after a previously set time limit, which would prevent further learning after this time, but we do not do this. The network weightings are never completely frozen so as to disallow learning.

However, with a settled environment, after a time Bashful's characteristics are fairly steady and it becomes difficult to move him from his habits. Nevertheless, some further learning is always possible, and this is vitally important. If the characteristics of the environment are changed slowly, it means that Bashful can gradually adapt his strategies in line with the changes. Secondly, if a dramatic change in strategy is suddenly required, Bashful will adapt rather slowly since he takes some time to get out of the habits to which he has become accustomed. Therefore, to an extent he still tries to behave in the way previously learned, even though he is now told regularly that it is no longer appropriate.

One feature of Bashful's behaviour led to a new design in the robots. In some cases he learned to spin to the right when he encountered an obstacle in front and to the right. When he did this the obstacle, perhaps a wall, would then be detected in front and to the left and he would spin to the left. This would put the obstacle on his right again, causing him to repeat the sequence. In this way he would occasionally learn to approach a wall and then to shake his head from side to side in front of the wall. It was thought initially that Bashful was learning a rather silly characteristic. However, on investigation it was found that he was in fact learning a perfectly sensible technique, but that his physical design was creating the problem.

The first generation of Seven Dwarfs, in common with Bashful, simply had a face with ultrasonic eyes to the right and left. Because of this design it is quite possible, when the right eye is pointing directly at a wall, that if the robot spins to the right his right eye moves away from the wall but his left eye moves closer to it. The converse is, of course, also true. An improved Bashful was, therefore, designed with a central, forward-pointing ultrasonic eye, as well as the original eyes to the left and right. This approach immediately removed the head-shaking problem, although it added some complexity to the artificial neural network. It is interesting to compare the new central eye system with insects, which only have eyes to the front left and front right. Such eyes are, however, compound, with lots of small segments, many of which are actually pointing forwards. So, in common with insects, Bashful needs to be able to sense directly in front of himself as well as to the right and left.

The number of possible situations depicted by the ultrasonic eyes and the number of potential actions are in reality much greater than have so far been mentioned in this chapter. Not only does the central eye add complexity but each eye can have much finer information to indicate the distance of an object. Also, the wheels can go at many different speeds, both forwards and backwards. All of this means that, even for this relatively simple example, there can be many more values involved in the robot's artificial neural network and the weight learning that is associated with it.

As further sensors are added, with corresponding actions of a different kind, so even more complexity is included. Infrared receivers provide a different type of information which requires that particular signals are learned. The response to such signals could be regarded as basic instincts, in terms of chasing after food or hastening away from a predator. However, the particular signals which relate to food or predator still need to be learned. This can be done either by extending or adding another layer to the neural network. In this way one layer of the artificial neural network deals with avoiding obstacles whereas the next (higher) layer indicates when food is nearby or a predator is close, and that a more important action is necessary.

The action of Bashful's brain has been described thus far in human terms with regard to the distance of objects from his ultrasonic eyes and the direction and speed of his wheels. It is acceptable for us to look at his brain in this way because then we can gain an understanding of what is actually going on. In reality, however, Bashful sorts himself out. His brain takes in signals from his ultrasonic eyes, categorises these signals and then causes an appropriate action with an aim to satisfy goals which are internal to Bashful himself. In this way Bashful arranges the operation of his brain in a way that best satisfies himself, given that we have provided his overall goals.

Bashful learns associations and respective actions which are appropriate to himself, his sensors and his wheels. As more sensors are added, so more associations are necessary, but the total remains appropriate to the sensors he employs. In the same way the responses of other mammals are appropriate to the sensors they have and actions they can carry out. In particular, the associations and respective actions which we humans learn are very much appropriate to our actions. When we make use of other sensors we still convert these into signals which are useful to the sensors we already have, before we process them with our brain. As an example, radar information is usually passed visually from a flat monitor screen to our brain through our eyes. An alternative would be to feed representations of the radar signals directly to our brain, which would then have to learn how to deal with them. When further sensors are added to one of the Seven Dwarf robots they directly affect the robot's brain.

Earlier I said that Bashful's brain was equivalent to 40 neurons, which is considerably less than the 100 billion found in a typical human. Despite this, many features, such as learning, can be witnessed and different behaviours and characteristics can be seen. Only fairly low computer processing power is required. In fact the original Bashful employed a Z80 microprocessor, which is a fairly old and relatively simple technology. Even with this limitation Bashful still has spare brain capacity. To put it bluntly, it is sufficient to balance processing power with the tasks to be carried out. In many cases humans use computer machines which contain much more processing power than is actually needed for the job they are doing.

To put Bashful's brain power in perspective, some worms and slugs are reported to have approximately 20 neurons in their brain, whereas honey bees typically have something like 100,000 neurons. So Bashful, although down the scale, is ideal for studying behaviour and learning. When watching him one quickly realises that there are many parallels with the animal world.

It is worth pointing out that the cybernetics department actually has quite a number of Bashfuls, a generation of them, in fact. The original Bashful, which appeared on *Blue Peter*, was really the first of the second generation of the robot Seven Dwarfs, in which learning is included. The later versions also have the third ultrasonic eye and infrared receivers for 'food' location and station docking.

The third generation of Seven Dwarfs is easier to describe in that they were put together *en bloc* and have one simple technological advance on the previous two generations: they can communicate with each other.

The robots can each not only receive infrared signals but can also transmit their own, which are unique in that they are at a different frequency for each Dwarf. In addition, each robot can make slight variations to the main signal in order to pass messages to other robots. The main signal is known as the carrier signal, because it is there to carry messages. The message signal is known as the modulating signal as, to transmit messages, it modulates or changes the carrier signal slightly. Each robot can, therefore, be uniquely identified through the infrared signals it emits.

The robots in the third generation have been given a pro-grammed basic instinct: when a number of infrared signals, at different frequencies, are received, each robot will move towards the direction from which most signals are coming. If only one such signal is received then a robot will move towards that signal. It is worth mentioning that such a response could easily be learned rather than provided as a basic instinct. All that would be required is some form of reward and/or penalty to indicate good or bad behaviour, thereby strengthening or weakening neuron weightings. As an example, the reward could be an evolved, instinctive, internal positive indication or it could be an external physical treat, such as the passing of some information. These can be directly likened to pleasure or a payment, just as a penalty can be likened to pain. The key purpose of the third generation was, however, to show communication between robots and emergent group behaviour. Learning group behaviour can come later.

When a number of the third-generation Seven Dwarf robots are in a corral at the same time, each robot checks for incoming infrared signals from other robots. It does this by repetitively polling around the different frequencies – that is, inspecting each of the relevant frequencies in turn – noting which robots are signalling. It is very much like a human listening for particular sounds. When a robot receives a number of signals, it moves towards the robots which are transmitting them. So, if all of the robots in the corral are transmitting a signal saying the human equivalent of 'Here I am', they will tend to move towards each other in response.

All the robots also have ultrasonic sensors, and with these, just like generations one and two, they can avoid hitting things, either through a program or through learning. So, when they move towards each other in the corral they will still also avoid hitting each other, the corral wall or other objects. The result is that they group together, staying a small distance apart, bobbing and weaving about, due primarily to physical differences and differences in signal strengths between them. If left like this, they would stay together in a close-knit pack, going nowhere. A further feature has therefore been added.

Each of the robots has been given the *talent* to be a leader, if it so decides. When a robot has decided that it wishes to be a leader it then sends out a different signal, that is, one at a different frequency. All the robots can recognise a leader signal when they 'hear' it, with the result that they follow the leader signal rather than heading towards the majority of robots. It is, however, quite possible for there to be more than one leader robot, in which case a follower robot will move either towards its nearest leader, whose signal it is receiving most strongly, or towards the leader who has most followers, because that is where most signals are coming from.

But how does a robot actually decide to be a leader? At present it is quite simple. Ordinarily when the robots group together, and the bobbing and weaving takes place, each robot will generally see several other robots in its vicinity. From time to time, however, a robot will find itself with no other robots immediately in sight. Its natural tendency (basic instinct) is then to move forwards, which it duly does, at the same time laying claim to be a leader by sending out a leader signal. Once the other robots receive this signal they then stop their bobbing and weaving and chase after the leader. It can be argued, therefore, that one of the major factors which makes one robot a leader among robots is its refusal to listen to others. Several leaders can emerge if three or four robots are looking in different directions at roughly the same time and none of them can see any other robots. Each will then decide that it is a leader and will head off in the direction it sees fit, with complete disregard for any of the others.

A leader will remain as a leader until it is faced with a number of other robots in front of it. At this point it loses its leadership status and returns to the ranks. It is as if the leader, when faced with a number of robots coming in the opposite direction, says, 'Hey, I don't want to be a leader any more; let someone else do it.' Such a situation can occur when the leader of one group comes into contact with another group. Alternatively, when a leader guides its group into a corner, on spinning round to come out of the corner it faces its followers chasing into the corner after it. Any sensible robot, in this case, would wish to get rid of the leader's job straightaway, which is what happens.

It is quite possible that when a robot decides it is going to be a leader and heads off in one direction, no other robots will follow it. The other robots could easily all be following another leader robot in a completely different direction. It can also occur that a particular robot moves too strongly/quickly when leading, which can cause the other robots simply to abandon it and let it head off on its own. The solitary robot will then continue on its own until it turns and sees the other robots elsewhere, at which point it gives up its leadership pretensions and rejoins the pack. Being a well-supported leader is very much a case of a robot choosing the right moment to become a leader and then heading off in the right direction (no link with politics intended!).

Once a group has become leaderless then they will either become aware of a leader of another group and, as a group, follow after it, or conversely, if no other leader is in the vicinity, they will bob and weave until a leader emerges from their ranks. This could be the same robot which previously led the group, or it could be a new leader.

The end result of these characteristics is that the robots behave very much like sheep. They flock together as a group, moving around close to each other in an area without actually colliding with each other. After a while a leader takes them off somewhere else, where they flock together again for a while, and so on. More than one group or flock of robots can emerge. However, that is usually short-lived, as one group will soon see another and consequently the several groups will merge into one.

All of the third-generation Seven Dwarfs have a range of features coupled with different sensors. They move around in an area without touching anything, something which they can learn to do. They can also locate a 'food' station and charge their batteries. They can communicate with other robots, of which they are aware, and exhibit group behaviour, following a leader. Several of these characteristics are complementary. However, others – for example, flocking whilst avoiding obstacles – are contradictory in that each robot has a tendency to move towards another robot and yet also has a tendency to move away from another robot!

What causes a particular leader to emerge? There is certainly the aspect of a robot being in the right place at the right time. However, the physical attributes of the robots also make them more or less likely to be a leader. A robot that does not 'listen' well to the other robots has a better chance, as indeed has a robot that is physically more powerful than the others. In this latter case the robot can move around more quickly, perhaps because its battery is more fully charged or its wheels move more freely, and in this way it is more likely to get into a position whereby it becomes a leader. In witnessing them over a period of time, one or two of the robots tend to be leaders quite often, whereas one or two others are rarely given the opportunity, and indeed when it does occur it may be rather short-lived.

The method of selecting a leader appears to be natural for the robots, although parallels can readily be drawn with animal groupings, and even humans. When a powerful leader emerges and moves off swiftly, less capable robots can easily get left behind and can even fall off the pace. Conversely, a weaker leader causes severe bunching, with the more powerful followers being held back. It is quite possible for other leadership selection procedures to be imposed on the robots, although it is felt that none of these actually matches up to the procedure in operation.

One possibility is a dictatorship, in that by some means one robot can be selected as a leader and will subsequently always remain so. This is not so exciting to watch scientifically, as the following group simply goes along in the mode of the dictator. If he is powerful, then weaker robots are always struggling to keep up, whereas if he is weak the more powerful robots are severely hampered. Much the same is true of a 'hereditary' leader. An alternative is for the robots simply to take turns at being the leader, each one leading for a set time period.

This is more interesting, but again falls foul of the extremist leaders.

The most viable alternative is for a democratic selection to take place, but such a procedure would require more brain power than 40 neurons. Firstly, each robot would need to make a judgement based on a number of measured factors, and would need memory so as not to forget what has previously happened. Although an interesting concept, and technically quite possible, democratic robot elections are beyond the present robots' brain capabilities. What is possible, however, is that the robots standing for election could emit a number of signals indicating how they would behave as leader; for example, how fast or slow they would go. Each of the other robots would then have to decide how much they believed those standing for election. The strength of the goal, to be elected, would then influence how much those that were standing would exaggerate claims of their future performance.

At present a leader ends his reign when he sees quite a number of robots blocking the direction in which he wants to go. This, again, seems an appropriate way to go about things. It could be, though, that one leader does not give up until a challenge appears in the form of a newly emerged leader. A selection or battle between the leaders would then have to take place. Elections, as discussed, are one possibility. However, in theory it would also be possible for the leader robots to battle it out between themselves until only one of them is still physically capable of leading. For example, the feature of avoiding obstacles could be overridden at the time of a battle so that challenging robots hurtle towards each other as fast as they can, retire a distance, and then repeat until one remains. Although such a characteristic is easy to instil in the robots, it would be an expensive way of experimenting and is not preferred to the method we have in operation.

When studying a single robot, certain behaviours and characteristics can be directly compared with those of insects and animals, including humans. Witnessing a very simple response of a robot makes one think again about a similar behaviour in a human. With the third generation of the Seven Dwarf robots, group behaviour can also be investigated, with leadership roles being of particular interest.

On one occasion I took the robots to demonstrate at a girls' school in London. The hall in which the demonstration took place was rather cold. As a result, one of the robots was extremely slow, and did not seem to want to move. The other robots grouped around the unfortunate individual, strutting backwards and forwards in a provocative manner. Much sympathy for the poor robot emanated from the girls watching, and from myself as well, for that matter. The robot was trapped, was being bullied by the others, or so it appeared, and looked desperately frightened. When I rescued the individual from his predicament I was cheered loudly.

The third generation of Seven Dwarfs can communicate between themselves, and this opens up a whole new area of research. What do they communicate to each other, how are they affected by this and how do they behave in response? What are the plans ahead for the robotics program and where is this likely to take us?

Chapter Eleven

What Next With the Robots?

To describe the characteristics and behaviour of the Reading Seven Dwarf and walking robots in terms of artificial intelligence is perhaps something of a misnomer. No attempt is being made to replicate artificially behaviour that would require intelligence if exhibited by a human. Each of the robot's behaviours is appropriate for that robot and its capabilities. However, in some instances comparisons can be drawn between the robots and insects. At other times similarities can be seen with some mammals, and on occasion even with humans. For the most part, though, the robots currently exhibit a very low level of intelligence, so much so that Dave Keating has called it 'artificial stupidity'. This may, or may not, be a more appropriate description. Perhaps you can draw your own conclusions at the end of this chapter, when I have described what is happening with our robotics programme now, what is likely to happen in the near future, and what other possibilities there could be.

So, what is going on at the moment? Two researchers, Ian Kelly and Iain Goodhew, have been driving the Seven Dwarf programme forward. They were responsible for building the third generation of communicating robots, as well as introducing some of the other features, such as predator/prey behaviour and docking at a recharging station. Ian Kelly's latest task has been to set up teaching and learning between the robots.

Some of the robots are now also fitted with a radio link, which allows them to communicate fairly detailed information to a remote station, such as a computer, and also to each other. Taking the second of these points first, robots are now mutually able to learn. This means that, for example, two robots can each have an independent corral in which to learn, and each one can start to learn how to move around and avoid hitting things. However, rather than simply learning alone, as described in the previous chapter, robots can pass on to each other their experiences, successes and failures alike by means of radio communication. So, taking one Seven Dwarf robot, it not only learns from its own experiences but is kept up to date with the experiences of at least one other robot at the same time. Interestingly, arranging for two robots to learn mutually is fairly straightforward, whereas with three or more, problems arise as to which robot is to be believed or which one should be treated more seriously.

Two robots, mutually learning how to move around in their own individual corrals, can result in quicker learning. In simple terms each robot is itself moving around, and therefore obtains its own trial-and-error information, but at the same time it also receives trial-and-error information from another robot.

However, the robots do not have to start learning at exactly the same time. One of the robots can learn on its own in one corral, and once its behaviour pattern appears to have settled down, another robot can be started off in another corral. The first robot then passes, by radio link, very strongly defined information to the second, 'younger' robot, along the lines of 'These are the situations you will face' and 'This is what I do successfully in those situations.' Rather than mutual learning, the relationship has clearly become one of teacher–pupil or parent–child. A difficult question is then: How much weighting does a pupil robot put on its own experiences, as opposed to what the teacher is telling it? Also, does the teacher put any weight on what the student experiences? The student could possibly come up with a new way of dealing with a particular situation. Should the teacher take any notice? Clearly there is a sliding scale, ranging from a pupil putting confidence only in its own experiences and not relying on the teacher at all, through a balanced approach based on the experiences of both, to the other extreme of a pupil relying completely on what the teacher says, putting no confidence in its own experiences.

When a pupil gives no weight at all to the teacher's information, it is learning completely independently and will take longer to experience the range of possibilities. It will also take longer before the behaviours of the robot have settled down. As it is independent, it is likely to learn different characteristics, different ways of dealing with things from the teacher's. As an example, the teacher may have learned to behave in a certain way when confronted with a corner. If the pupil takes no notice of the teacher, it is likely to learn a completely different behaviour when faced with the same situation.

On the other hand, if the pupil only believes what the teacher tells it, and places no emphasis on its own experiences, then it could have problems if its own environment is substantially different from that of the teacher. If the pupil's corral has corners and the teacher's does not, and if the pupil does not allow itself to learn from its own experience at all, then it will simply have to go on the best that the teacher's behaviour can offer, which may be very poor. But on the positive side, as the teacher's brain is well settled, its mode of operation can be transferred *en bloc* to the pupil; that is, a brain dump can occur. The pupil's brain can immediately become a copy or clone of the teacher's.

But what is actually happening when the teacher robot downloads its brain operation in its entirety into the pupil robot, so that the pupil will behave in exactly the same way as the teacher? Quite simply, the teacher is programming the pupil. One robot is programming another. There is, of course, nothing to stop a domino effect developing: the pupil robot itself programmes yet another robot, and so on.

The two extreme cases – that is, the pupil robot taking in either everything or nothing from the teacher robot – are relatively easy to set up. But humans operate somewhere between these two extremes, and so this is what we have done with the robots. A pupil robot places a balanced weighting on what the teacher tells it and what it experiences itself. The actual balance of the weighting is currently a research topic.

The situation is, however, slightly different from the human case, in that the teacher and pupil robots are both moving around in their respective corrals. The teacher passes on information on the situations it encounters, as it encounters them, along with information on the already successful strategies it uses in those situations. The pupil, at the same time, is probably facing different situations and is gaining experience from its own trial-and-error testing. So in some situations that the pupil robot faces early on in life it will have no prior information from the teacher, whereas when encountering other situations, the teacher will already have passed on details of what it does in those cases. At present the teacher robot does not lead the pupil robot to particular situations in order to tell it what to do in those cases, which is perhaps a more human teacher–pupil relationship.

So far we have experimented with different pupil belief factors. That is, how much the pupil believes from its own experiences in relation to how much it believes from what the teacher robot tells it. The problem is, however, complex. When a pupil robot is passed a lot of supporting information from the teacher on what to do in a situation the pupil has not yet faced, how much final adapting is left until the pupil actually encounters the problem itself? In other words, should the pupil completely make up its mind on a situation before it has itself experienced that situation?

Another difficult decision arises when a pupil robot experiences a situation to which it finds a successful solution, only to be informed subsequently by the teacher robot of its different and yet also successful solution. Averaging the two solutions may produce an unsuccessful strategy, so which solution should the pupil robot choose, its own or the teacher's? Which could be better?

In facing problems such as these with the robots, it is extremely difficult not to draw parallels with human or other animal behaviours. But why should we not? Thinking about what we do as humans often provides a direct answer to the problem with robots. It is important to realise that such problems are largely the same whether they be in robots or animals, although there may be some differences in terms of species-dependent characteristics. For example, if a human teacher smiles, the human pupil may be more likely to believe what he is being told. Is this not just being a good salesman? It is, however, difficult to relate such a feature directly to the robots, as at present they do not have the physical capabilities to smile.

So what would life be like if you yourself were a pupil Seven

Dwarf robot? What would it actually be like to be of small size with fairly limited brain power? Perhaps it is best to start by thinking what life would be like as a worm or a small insect. Goals in life are relatively simple, but the brain power and physical attributes to deal with those goals are also relatively simple. Is there any chance for abstract thought? Are you conscious? These questions go back to Chapter 5, and we may never have a clear answer to them. Here I would like to concentrate on what we do know.

As a pupil dwarf, until you are switched on, there is nothingness. But once you are switched on you come alive. Your life is, therefore, dependent on someone switching you on and off. The only sensory systems you have are ultrasonics, infrared and radio. You have no human eyes, no human ears or nose, so the only information you receive on the world around you is through the three senses you do have. A basic instinct tells you to move around and another tells you it is not good to get close to other things.

You try moving around and are told it is good that you have not hit anything. So you are positive in what you are doing. Then you hit something and are told that that is bad. So you are not so positive about doing the same again. At the same time, through your radio connection, you are being told that when you face a particular situation it is good to do a specific thing. So it makes you more positive about doing that specific thing, even if you have not yet faced the particular situation. All the time you are moving around, being more and more positive when you do not hit anything, until you get to the stage where you are moving around and never hit anything unless occasionally you make a mistake, which is acceptable. After all, you are a machine!

Then there is the infrared signalling, and you have a basic instinct to send out your own signal. At the same time you can pick up a number of infrared signals from a distance, so you move towards them. But when you get near, you do not move too close because you have learned not to hit anything. Suddenly you pick up a leader signal and all your attention is focused on following that signal. You do this for a while until the signal disappears, and then you return to staying close to where there are a lot of other infrared signals. Remember, you cannot see or touch the other robots; you just pick up and respond to infrared and ultrasonic signals, although your radio link could receive some information at any time.

As you move around, suddenly you find that you cannot pick up any ultrasonic signals any more. There is nothing in front of you, no barrier, nothing to get in your way. So you send out your own 'I am a leader' signal, and head off in the direction of an open space. After moving around for a while you eventually pick up some ultrasonic signals, either from a fixed object or from another robot, and as a result change your own signal back to its original form. And that is your life. This is what you do and this is all that you do until you are switched off or your battery goes flat or, if you have the facility, you detect your battery is running low and so go off to charge yourself at a charging station. There is, at present, no mid-life crisis for you as a Seven Dwarf robot. Once you have learned how to move around, then essentially that is all you do until you die, that is, until you are switched off. However, that is about all your brain can cope with.

So life as a Seven Dwarf robot, when compared to life as a human, may not be particularly exciting at the moment. However, it is changing as different features are added. One recent addition, brought about chiefly by Ian and Iain (Kelly and Goodhew), was to cause a remote computer to pick up the radio signals being transmitted by a robot. In this way it is possible to see, on a computer screen, what a robot is trying to do, why it is doing it and just what the result is. So it is easy to see when a robot has, effectively, learned all it is going to learn and is simply moving around with a fairly fixed brain pattern.

The radio link between robot and computer is two-way, which means that a computer can act as a teacher robot and pass information to a pupil robot. By coupling two computers together it is thus possible for a teacher robot in one corral to pass on its experiences to a computer, for this computer to transmit the information to a second computer and finally for the second computer to send the information by radio to a pupil robot in a second corral. The pupil robot can then learn, partly through its own experiences and partly from the teacher robot via the computer-to-computer link. Such an experiment was first conducted successfully in the cybernetics laboratories at Reading University in spring 1996.

The computer-to-computer link opened up a whole new aspect of research. Via the Internet, computers can be linked up together around the world, with messages being passed from one computer to another in a matter of seconds. This allowed us the possibility of a teacher robot, situated in Reading, passing what it had learned to pupil robots located in different places around the world. Such an experiment was successfully conducted for the first time in November 1996, with pupil robots located in New York and Tokyo being taught by the Reading teacher robot.

The success of the experiment was crucial in proving the power that machines can possess. The experiment showed that one machine, located anywhere in the world, can learn to behave in a certain way, based on what it is taught by another machine. Subsequently this machine can, in the space of a few seconds, teach other machines to behave in roughly the same way, given that they will each have their individual characteristics. The whole foundation of what the first machine learns is on the basis of what it believes to be a positive action. It could be programmed by some human with what is positive and what is not, but it could also learn certain 'positive' actions itself, such as a behaviour which causes it not to be switched off. But the robots at Reading do not have such a power. All that we have done is to use them to show just what is possible.

There are, however, other features which are being incorporated in the Seven Dwarf robots, although to accommodate them it has been necessary to increase the microprocessor-based brain power. The latest versions of the Seven Dwarfs have brains which use the equivalent of about 500 neurons. Iain Goodhew has been responsible for putting together a compound eye which sits on top of a robot. The eye consists of a large number of phototransistors, each pointing in a slightly different direction. Essentially, phototransistors respond to the presence or absence of light. As the light intensity picked up by a phototransistor changes, so does the electrical current which each phototransistor allows to flow.

The compound eye, which is somewhat like the eye of a fly,

is connected directly to an artificial neural network consisting of a number of neurons known as Minchinton cells,[56] named after their inventor. These neurons work especially well on light patterns. With the robot placed in a particular position, for example, in a room, that position will have a light pattern associated with it. This pattern will, of course, depend on whether electric lights are switched on or off, where windows are positioned, and so on. The robot is then taught to recognise the light pattern at that position, the Minchinton neurons being organised to remember roughly the light pattern being picked up by the eye. I say 'roughly' because they operate, rather like our own eyes, by learning a general rather than an exact pattern. Subsequently, when the neural network sees a similar light pattern, it recognises it.

The robot with its compound eye can then be placed elsewhere in the room, and asked to find its way back to its original position. It can do this by seeing, from wherever it is, where the light gives a better match than at present to the light pattern that it was originally taught.

By continually moving in the direction which gives a better match, the robot is likely to find its way back to its original location. However, using only the light intensity as a guide can cause problems, particularly when a room uses strip lighting or has repeating light patterns through windows. In this case the robot may home in on a position which gives a relatively good match, but which is not the original position. Moving in any direction from that point gives a match which, though the best in that locality, is not the best within the whole room. Therefore, after the robot gets to a local optimum and the match is still not very good, the robot heads off in a random direction for a short period and tries again.

By using the compound eye, it is easy for the robot to be taught, and then later to recognise, a position where either strong or distinctive lighting illuminates the area. In particular, if a bright light is shone at a particular, unusual angle, this is easy for the robot to home in on later. However, it is where ordinary, everyday lighting is used that the problem becomes more difficult, in that no simple and easy distinguishing features may be present.

Learning positions by means of a compound eye can easily be extended so that the robot can learn to recognise a room in terms of a light intensity map. The robot can then find its way around a room very accurately by means of its light intensity guide. This, however, is something for the future. It has not been done yet.

Another ongoing project is to add a number of facilities to the robots. Initially a small two-fingered gripping device with an up–down wrist action is being added, together with a touch sensor on the gripper and an audio (sound) sensor to detect slippage. The aim is to incorporate the gripping device, along with its related sensors, into the robot neural network structure.

With the gripper attachment a robot will be able to interact with, pick up and move objects in its environment. In terms of Brooksian levels,[53] as described in Chapter 9, this is going some way to tackling levels 6 and 7. At the moment a single two-fingered device, something like a crab's claw, is complicated enough for one of these small robots to deal with. Having more fingers would be problematic, as I indicated in Chapter 4, and to introduce two grippers and, therefore, face a coordination problem, would make things much more difficult. However, neither of these things is necessary at the moment. The overall aim is that a robot will be able to learn about and control each device and sensor with its neural network. It is not intended that complex mathematical equations will be derived in order that the control can be carried out.

An interesting aspect of research is being carried out by another research student, Ben Hutt, and that is evolutionary robot design by means of genetic engineering. Essentially, the features and characteristics of a number of robots can be entered into a computer, along with details of the environment in which they exist. Important factors are how they obtain power, whether any predators exist and general survival requirements. Each robot can be investigated to see how well it copes with the conditions in its environment. Features of robots are then intermingled genetically to produce a new generation of robots, purely simulated within the computer. The process can be repeated with the new generation and so on, through many generations.

A big advantage of genetically evolving robots through generations within a computer is that it can be carried out in a matter of minutes. This compares with the hundreds or thousands of years it takes for features to evolve in humans or other mammals. So from a set of real robots in a real environment, features of the robots can be genetically engineered within the computer for many generations, until a robot is evolved which is much better able to cope within the environment. This robot, or even a complete new generation of robots, can then be physically constructed.

Rather than a few relatively small steps taking place, as is the case with mammalian evolution, with computer-based evolution many physical changes can occur between one physically constructed generation of robots and the next. The intermediate generations merely exist within the computer and not as real-world entities. Apart from the obvious genetic ploy of mixing features (genes) from the better robots in a generation to produce improved offspring, it is also possible to try mutations. In this way new features can be added or existing features changed in a fairly arbitrary way. If a mutation is successful then that robot will thrive within the simulated environment, and will itself have an influence on its offspring in future generations. It may well be that an original mutation will, if it is reasonably successful, further evolve through many more generations. Conversely, if a mutation does not help a robot in one generation and it does worse in the simulated environment, then that robot can simply be abandoned and will take no further part in the simulated proceedings.

Mutations can be used within the simulations to try out new ideas to determine whether the robots can perform in a better way. If the idea is good, then it will evolve over a number of generations into something even better. If it is a bad idea it will have a fairly limited effect and will quickly disappear, although in some rare cases two bad ideas can combine to produce a good end result, so care must be taken not to scrap a bad mutation too quickly.

The whole subject area of applying genetics to engineering problems, not just in improving robots, is a relatively new field and has opened up new research directions. A number of factors are of critical importance. One of these is the size of the population of robots in a simulated generation. When the next generation is produced, should the population be allowed to grow or should it be held at the same size? Keeping it the same size means that only a constant number of robots can be retained from generation to generation, and if we are looking for improvement then some obsolescent robots must be scrapped. They could, however, have been useful for a long time to come. Conversely, allowing the population to grow and grow not only takes up a lot more computer time when searching through to see which robots best cope with the environment, but also the average level of performance can easily be dragged down by a number of poor mutations.

When evolving robots genetically it is important to be able to record the salient features of the environment in a relatively concise form on the computer. The same is of course true about the features of each robot. Imagine trying to write down the important features of life on Earth which have affected human evolution, and then subsequently to record the important, relevant features of each human which affect evolution and which make us better suited to our environment. It is extremely difficult to say immediately which humans cope better with life on Earth, and why, and so it is difficult to categorise their salient features within a computer. Yet humans evolve genetically in the real world. If the environment changes – for example, global warming occurs – then those who are better able to cope with this will produce offspring who are even more able to cope with it, and so on.

Being able to represent concisely both the environment and the creatures within it – in this case robots – in terms of computer code is thus a very difficult problem. No matter how many subsequent generations of robots are produced within the computer, they will only be considered in terms of their performance in the computer-represented environment. If, after a number of generations, a super-robot appears which copes amazingly well in the computer environment, then just how well a real-world version of it copes in the real environment is dependent on how well the computer-simulated environment represents the real thing.

An interesting feature of the evolutionary robotics research is the number of robots which can be brought together to produce an offspring. The genes of several robots can be intermingled to produce a robot of the next generation. Ben Hutt calls this 'cyber sex'. However, at present it is not known whether this is a good feature or a bad one. A friend of mine at British Telecom, Peter Cochrane, has pointed to the fact that in the real world, mammals seem to have converged on two as being the appropriate number for sexual reproduction, so perhaps there is something good in this number. However, I am not so sure that such a restriction need apply to simulated sexual reproduction, particularly between robots, unless some clear reasons can be seen for restricting it in this way.

As well as progressing work in evolving robotics we have also been pushing ahead with our programme on walking hexapod robots. Rak Patel, a researcher in the cybernetics department, was largely responsible for getting Walter up and walking, and Darren Wenn, another researcher, has been chiefly responsible for Elma.

The basic design and abilities of Elma were described earlier in Chapter 9, however, we will now look at the most up to date research taking place with her and the plans we have for her future.

The original idea for Elma was that she would learn to walk by trial and error, and this occurred for the first time in March 1997. Most walking robots at this time work on a system of signals which are sent to the motors or hydraulics which operate each leg. Walter is a step on from this in that each of his legs has its own local intelligence, by which it looks after itself. Nevertheless, both the individual leg movements are the coordinated leg actions which cause Walter to walk, are fixed. Elma however, learns how to coordinate her legs without falling over, by trying different individual leg movements. At first she tends to stumble, occasionally bumping her head or bottom on the ground, but after about 15 minutes, her legs are organised in a gait which allows her to move around successfully. If after this she makes a mistake and slips, it is more than likely due to her trying to navigate around an obstacle, or such like, rather than any fundamental errors in her learning.

The sensors in Elma's feet detect when each leg is pushing against the ground; this aids the assimilation process as it should be possible, through this, for her to learn to cope with uneven ground and oddly shaped objects. It will be interesting to see how easy it is for her to keep her body off the ground across rough terrain. The way in which she learns to walk can be researched, and as an example results from this show that a quicker learning routine often means a successful, but poor, final gait whereas a more laboured approach will produce a steadier end result.

It is interesting that with all the robots, once some form of learning is involved, in a particular situation they sometimes learn to do what one would expect from our knowledge of the animal and insect world, but at other times they learn to do things in a way that we have not expected. Because at this stage each robot's neural network and the related learning process is not too difficult to analyse, we can often come up with a plausible reason for the robot learning a particular characteristic. There is generally a direct link between a robot's physical attributes and what it has learned. As humans we tend to look at a problem in terms of human values and to view the situation through human sensors, such as eyes and nose. But on looking at a problem in terms of how the robot can sense things, what its goals are and what decisions it can make, it can become immediately obvious why the robot has learned a specific action or response.

By studying the way a robot learns things, what is and is not important in the learning process and what information it requires in order to learn certain characteristics can become apparent. This can have an immediate impact on how we view behaviours in humans, other mammals and insects. With the robots we are, at present, able arbitrarily to modify the learning process, to provide a more or less exciting environment, to add sensors or more information, or, perhaps most important of all, to witness a whole robot learning process in the space of a few minutes. From this a fascinating new field of research, known *as machine psychology* has opened up, in which not only can we have fairly strict control over the subjects and their environment but we can also open their brains up to try and figure out why they behave in a certain way. One immediate observation from the studies we have carried out so far is that many behaviours in us all, machines and non-machines alike, appear to be very basic responses. Often a simple survival instinct is present, or a fundamental reward such as food or pleasure is the driving force. In our robots, these instincts are manifested as the avoidance of predators or the need for a recharged battery. Also, it may well be that the more basic response overrides what would appear to be much more rewarding alternatives. As an example, after two weeks alone in a hot desert, what would you *immediately* select, given the choice of one gallon of cool drinking water, £10 million or your very own time-travelling machine? Certainly, our investigations so far with machine psychology have taught us to look closely, in any situation, at what is the most important underlying, basic feedback in action. Pavlov's dogs drooled because when the bell rang they thought food was on its way, not, I feel, because they thoroughly enjoyed the occasional musical interludes with bells.

I watched a television programme recently, in which the discussion was about children learning to understand. A three- or four-year-old child was shown a Smarties packet and was asked, 'What is in the packet?' The packet was opened to reveal not Smarties but pencils. The child was asked what they were and replied, 'Pencils.' The child was then asked, 'What did you expect to see before the packet was opened?' To this the child replied, 'Pencils.' The conclusion drawn on the programme was that the child did not yet have a good enough grasp of language to understand the question, and therefore gave a *wrong* answer. My own belief is that the response was much more basic than language understanding, and was the child's straightforward best guess at what the adult wanted to hear. Essentially the child was looking for a reward in the form of a 'yes' or 'well done'. With about ten pencils lying on the table and the word 'pencils' having been mentioned several times, if the questioner had instead asked, 'Who is the President of the United States of America?' the child would probably have had a stab at 'Pencils.' But is not this something akin to Pavlov and his dogs again?

It is interesting with the robots to look at how individual

247

characteristics affect the way in which one robot treats another. I mentioned earlier about the bullying that appeared to go on from a group of the Seven Dwarfs when one of them seemed to be weak because of being in a cold room. If subsequently the same robot became even weaker and effectively died by having no power to do anything, then it would have stopped sending out any signals at all. At this stage the other robots would have ignored it and just left it for dead. Yes, this analysis is full of human values. When considering that the robots were bullying the weak one, we could easily have viewed it instead as the robots caring for the weak one, rallying round and showing their support, thereby applying a different human value.

What actually happens is that the Seven Dwarfs flock together by means of infrared signals, a basic behaviour of theirs. The fact that one of them is a little 'under the weather' and, therefore, does not move much, means that the others tend to flock to where the weak one is, unless a leader takes them away, thus leaving the weak one behind. If no signal at all comes from this robot, then the other robots simply ignore it or treat it as any other object, such as a wall or a box. Essentially they know no better.

So can we relate this to human behaviour? Surely groups of humans do not leave other weak humans behind or completely ignore dead people? Well, if we cannot see a dead person lying in the road, how do we know that he is there? Was it not the philosopher Berkeley[57] who said something along those lines? The important thing is that as humans we must, in some way, sense the dead person. If we cannot see him, unless we discover him by means of another sense, possibly smell, then we do not know that he is there. Exactly the same is true for a robot. Unless it can, in some way, sense another robot which has expired, how does it know it is there? In this case the (now dead) robot was sending out infrared signals when it was alive, so perhaps if each robot had a larger brain with a better memory, it could still have some knowledge of a robot previously existing. It might then need some concept of alive or dead, perhaps in terms of whether or not it could detect an infrared signal.

There is no harm in looking at this behaviour and characteristics of machines in terms of human values, as long as we make judgements based on the physical and mental capabilities of each in its own way. A key feature of Berkeley's work is in his argument that the world exists only in being perceived by the mind. Each individual responds to the world which their brain can understand, through that person's senses. The real world to a person who is blind is most likely different to the real world to a person with sight. Moreover, Berkeley's philosophy is just as true for machines, other mammals and insects as it is for humans, and importantly so. 'The real world for a machine exists only in how it is perceived by that machine's brain.' This is an important diktat for all those who deal with computers, robots and other machines.

So, if one conclusion can be drawn from research into the behaviour of robots, it is that the results could easily help and have a dynamic effect on the behaviour of humans and other mammals. Coupling this with the fact that it is much less contentious to experiment with a robot and to meddle with its brain than it is to do so with a human, then viewing robots in terms of relative human values could be well worthwhile.

How individual robots learn or react to environmental features can be investigated to a greater extent than we have done so far: for instance, not only how much emphasis a pupil robot places on its own experiences as opposed to details from a teacher, but also how quickly and strongly the pupil believes its own experiences. As more features and abilities are given to the robots, so priorities need to be drawn between what, and how much, is being learned with regard to each of its abilities. Particular learning habits and traits can be investigated and comparisons made. One specific area of interest is how active the robot's learning process should be once an initial fairly rapid learning period has been completed.

As well as the characteristics of individual learning, it will be interesting to investigate how the Seven Dwarf robots can learn to respond to and interact with each other on the basis of each robot having its own separate identity. Currently, apart from individual physical characteristics which differ slightly, each robot simply sends out its own infrared signal. Two things, therefore, need to change. Firstly there need to be more physical differences between the robots, and this is being directly remedied in that grippers are being fitted, and compound eyes added. This latter addition is particularly useful as, at the same time, the robot's brain will also be physically different. Secondly, the robots need to be allowed much greater flexibility in their modes of communication.

Communication between individuals is extremely important, and in machines the potential is tremendous. Humans have relatively poor interhuman communication skills when compared to intermachine abilities. Indeed, we use machines to help our own communication. Allowing machines the ability to learn to communicate with each other is an interesting step, even if at first the medium in which they communicate and the range of communication allowed is restricted in order that a speedy solution can be obtained for experimental purposes.

Even by restricting our own robots to infrared communication some results should be achievable. In the first instance each robot can emit a range of signals while at the same time attempting to receive signals from other robots. On receipt of a signal a reply can be sent out, perhaps a particular frequency. A key aspect of the robot's learning to communicate is that, as with other types of learning, some reward is given to indicate when it is communicating in a good way.

With a robot pupil–teacher relationship it is easy to see that just as a pupil robot can learn how to move around without hitting things, so the same robot could learn, from the teacher, to send out simple signals which relate to the situation it is in and what it is doing about it. By means of a radio link, with which the teacher sends instructions, the teacher can also receive information from the pupil on its situation and its response. It is then possible for the pupil to send out an infrared signal corresponding to its state and for the teacher to give a response which says whether the signal should be of a higher or lower frequency, or is in fact the correct signal sequence. Once the communication abilities of the robots have been opened up, with each robot learning to communicate with others, it is exciting to look at the variety of ways in which the range of information communicated can be considerably expanded.

As a simple example, if a robot is fitted with a gripper, there is no reason why it cannot send out signals relating to whether its gripper is open or closed, or whether the object held is slipping, and so on. Indeed, in the first instance it can select its own signal to send and require other robots to learn what the particular signal means.

There is no reason why the other robots cannot do this, as long as they can make some link with a gripper opening or closing. Remember, from Berkeley's philosophy, the robots do not actually need to know that a gripper, or whatever, is being used. They merely need to perceive that it affects them in some way. There is also no reason why two or more robots cannot themselves learn to use particular signals to indicate something, even if it is something fairly straightforward in the first instance. In particular, when from time to time one robot becomes a leader, there is no reason why it cannot redefine signals, as long as each robot knows that a change is taking place, or introduce new signals, as long as the other robots are aware what these mean.

Bringing together the possibility of pupil robots learning how to communicate, with a leader's ability to change signals and bring in new ones with new meanings, has one major implication: there is no reason whatsoever why robot machines cannot develop their own language.

With the Seven Dwarf robots in the laboratory we can study the possibilities: we can strictly define the range of communication abilities, the type of signals sent and the flexibility allowed in learning and changing signals, the extent to which new signals are introduced and the context in which they are introduced; that is, what the signals mean. After leaving the robots to run loose for a while we would return to find that at first we do not understand what they are saying to each other. However, by closely watching them for a period, and having knowledge of the range of things about which they can communicate, then we humans would be able to learn what the robots are 'saying'. It would essentially be the same for us as learning a foreign language.

We can see with the small, friendly robots in our laboratory that it is physically quite possible for them to learn to communicate in a language which we humans do not initially understand. There is no problem here, as we know everything about which they communicate and the range and type of communication signals they use. In general, this would mean that it is quite possible in other circumstances, where such a tight grip is not being held, for machines to learn to communicate in a language we do not understand. But further, unless we know in detail the context, and the environment about which, and in which, they are communicating, then it will be extremely difficult for us to understand them. Essentially, with our small robots in the laboratory we can prove a point which could be very worrying.

So, what other avenues of research does the immediate future hold? One topic is that of 'unlearning', as distinct from forgetting. When one of the Seven Dwarf robots is learning how to move around the corral it will learn a number of ways of behaving in certain circumstances. In particular, when the robot learns a behaviour, quite strongly, early on in life it tends to retain that behaviour even though it may be a little strange or not particularly useful. The robot tends not to forget it. If we now look on this strange behaviour as something undesirable, then can it be unlearned, and if so how? If we can look into ways of achieving this with robots, then maybe the methods would also be useful for humans.

One unlearning routine is a straightforward brainwash; that is, just to reset the robot's brain to a pseudo-random arrangement. The robot then has to learn again from the beginning, and although in due course it will almost certainly not exhibit the earlier strange behaviour, it may well exhibit some other peculiar characteristics. With regard to humans, the concept of completely brainwashing a person is not only a little difficult to imagine in practice but also, I suspect, would not be overly popular.

However, if it is possible to locate, in the robot's brain, roughly where the neuron synapses relate to this characteristic, then the learning action can perhaps be restarted in just a small portion of the brain. Such an action would probably affect other parts of the robot brain to some extent, for a while at least, but in the long term the chances of learning a strange behaviour to replace the one that has been 'unlearned' would be minimal, unless, of course, identical environmental conditions occurred again, possibly in a particular sequence.

It is certainly well worth investigating the possibilities, particularly if an alternative can be found to some neurosurgery on humans. In the case of epilepsy, as was mentioned earlier, neurosurgery has produced limited results in effecting a cure. A large proportion of people with epilepsy can get by with drugs which carry out a suppressing act. For many more people drugs do not help, no matter how many are taken, and on occasion people die when in a fit. A large number of other neural problems occur, however.

My own father had agoraphobia, to the extent that he was not able to go to work. As his case got progressively worse it was eventually decided that a neurosurgical operation would be required, although a high risk was involved. I was nine years old at the time and cannot remember much of what went on, except that by operating on a specific part of my father's brain the surgeons at Smethwick Hospital were able to turn things round. After a fairly short convalescence my father was able to go outside in the normal way, although for a while he did have an extremely sharp temper. Initially this just meant that he occasionally stood outside the front of the house shouting at people for walking past with their hands in their pockets. Before long though he was able to return to work and resume life as normal.

If by studying in depth how artificial brains work, whether in robots or not, we can help humans who have neural problems, then it is well worthwhile, particularly if we can find alternatives to neurosurgery or the need to constantly load up with drugs in order to subdue a problem.

One other line of investigation researches into the way that a robot's artificial neural network is fitted together. As illustrated earlier, rather than throw an enormous neural network or huge computing power at a problem and hope to get something positive out of it, we use what appears to be a reasonable network size and type for each problem. When more features are added to the robots, extra pieces of brain are appended in order to deal with them. This means that for the Seven Dwarf robots their effective brains are taking on a fairly partitioned look.

We know that a human brain operates by means of different sections which largely deal with different features. However, those sections are all physically tightly coupled together. One aim with the Seven Dwarf robots will, therefore, be to try and achieve a synergy with the make-up and operation of a human brain from the bottom up, in a neural sense. So the build-up of a brain that is tightly connected, but also partitioned from an operational viewpoint, will need to be researched. My view is that, at the end of the day, a natural partitioning would be more effective than the enforced one we effectively have at present. In other words, we need to look to an overall robot brain which sorts out its own loose partitioning rather than having a stricter partitioning forced on it.

The whole problem of brain size and connectivity returns to the earlier discussion about bees and their collective powers. A hive of bees has a brain which has, in effect, many partitions, each bee's brain being a separate part of the total. But each bee brain is doing the same things and operating in a very similar way to every other bee brain. So although there is some gain in a lot of bees grouping together, it is rather limited because each of the small partitioned elements, each bee brain, is doing very similar things.

A human brain is bigger than a bee brain, but in the number of neurons is similar to millions of bee brains together. However, a human brain is acting as one, with roughly partitioned areas dealing with different things. If a human brain was partitioned up into thousands of small areas, each one operating in a relatively independent way and each one doing much the same thing, then that human would probably have the intellect of a hive of bees and not much more.

By looking at what is technically possible now, it is worth ending this chapter on a strong note of caution. At Reading we are nice, friendly people who are researching into intelligent machines. But what might be possible for people who are not as friendly and open as we are, who have a lot more money to spend than we do and who do not tell people what they are doing?

As I mentioned earlier, we have a 5 ft high Dalek-type robot that moves around on wheels. It is essentially a very big version of one of the Seven Dwarf robots. It has an infrared detector with which it can measure extreme heat, such as that caused by fires.

It also has a fire extinguisher nozzle which it points at a fire with the object of putting the fire out. The overall use of the robot is, therefore, very positive. It can roam around a building with no supervision, and instantly puts out any fire that it sees. It thus tackles what could be a dangerous job for humans.

But what *could* be done? What is technically possible now? Instead of detecting fires by means of an infrared sensor, the robot could be supplied with a camera-based vision system, perhaps as I described in Chapter 9. By connecting a neural network up to the vision system, the network could be trained to recognise smiling or angry faces, as was the case before, or perhaps to recognise people with glasses or blond hair, or even to recognise a particular person. Instead of a fire extinguisher the robot could have a machine-gun. It could roam around a building, not looking for fires, but rather looking for people wearing glasses or with blond hair or even a particular person. Perhaps as far as the robot is concerned grey hair is the same as blond. What does it do when it finds them? It shoots them, or at least it shoots a person whom it decides fits the description, or else it shoots a person whom it decides does *not* fit the description, as the case may be.

Is all of this purely speculative? I am sorry, but it is not so. It may be that no one would actually want to build such a being – at Reading we certainly would not – but technically it is possible, *now*.

Another worrying fact is that if the robot's objective in life was simply to find and destroy, then that is all it would do. It would keep searching until it found its quarry. It would, of course, have to go and charge its batteries from time to time, but that is not a problem; we know robots can do that. Also, the robot could easily learn its own target. It could learn that blond-haired people are its quarry, or that people, or anything that looks like a person, could try to switch it off and should therefore be hunted down. If a simple robot can learn to move towards a smiling face, then there is no problem for a robot with a machine-gun to learn to shoot people because they try to switch it off. All it would need to know is firstly, that being switched off is a negative thing and should be avoided at *all* costs, and secondly, what must physically happen for it to be switched off.

It does not need to be a single robot of this type; there is no reason why there could not be many of them. Perhaps they would even be communicating in a language we do not understand. But as they have wheels, surely we do not have to worry, because they cannot travel outside and they cannot climb stairs? Well, in a building the stairs must go somewhere, either up or down, and wherever they go, other, intercommunicating wheeled robots can be. Also, didn't I read somewhere about walking robots? Indeed, as we will see later, we can consider the whole building itself to be an intelligent robot machine.

But even though it is technically possible, surely no one would want to build a robot such as this. Why should they? Perhaps it is just about as likely as someone building a nuclear bomb which, when exploded, wipes out a whole city.

Chapter Twelve

A Fantastic Future?

Robot machines which have brains with about the same power as insect brains can do a number of things, in the same sort of way that insects can. The things such a machine does, though, are relevant to itself as a robot and are concerned with its own robot sensors and movements. In the same way an insect's brain deals with its own insect sensors and movements.

By connecting more neurons together in a higher density, with good connections between the neurons and relatively well-developed learning techniques, robot machines are becoming more and more intelligent. As previously mentioned, in the next few years we will have machines with an intelligence of about the same level as a cat's. Such machines will, therefore, require not only a well-connected brain, with many more neurons, but also some physical characteristics such as sight and touch as well as movement. In this way the physical and mental aspects of the machines can be well balanced. It may be that although such machines will do different things from cats we will be able to interact with them in the same way that we presently do with pets. Certainly, intelligent robot pets are a possibility.

A robot pet's view of the world would not be the same as that of a cat, because, most likely, its physical characteristics would be different. Importantly, one would imagine, it would think more in the way that we want it to than a cat does. We would be able to gene-program it to behave in certain ways. It might be difficult, as it is with real cats, to get robot pets to behave exactly as we want, however, because of their ability to think and act independently.

As things move on further with machine intelligence, so the

257

density of neurons will increase, as will the quality of connections between them and the associated methods of learning. Even thinking of robot dogs or cats which do not follow human accepted modes of behaviour is rather strange, such robots behaving more like wild animals than domesticated ones. But as robot machines become more intelligent still, can we ensure they are domesticated and follow important human rules? This is something that was addressed many years ago by Isaac Asimov. Asimov's three laws of robotics[58] are:

1. A robot may not injure a human being.*
2. A robot must obey the orders given by a human, unless this conflicts with law 1.
3. A robot must protect its own existence, unless this conflicts with laws 1 or 2.

Although Asimov's laws are purely fictional and are aimed at creating a society in which robots remain subservient to humans, they have been taken by some people, even now, as strict regulations to which robots must adhere. But one should be clear: they are just fictional ideals, nothing more and nothing less.

Coupled with the laws of robotics is a misconception some still hold that robot machines are all programmed by humans, in such a way that the above laws can be fixed into a robot's set of rules; then everybody can sleep easily at night.

It is certainly true that basic industrial production-line robot manipulators can be required to do one simple task repetitively, so they are programmed just to do that task. But as you have read in the chapters of this book thus far, robot machines are often far, far more than industrial manipulators. In particular, intelligent robot machines can learn and change their actions in accordance with what they have learned. This means that, dependent on what such machines learn and their physical capabilities,

---

* Asimov's first law in full is: 'A robot may not injure a human being or through inaction allow a human being to come to harm.' The law in this form is very easily self-contradictory in the real world, as will shortly be shown. The definition given above is, therefore, seen to be more realistic.

Asimov's three robotics laws are not realistically applicable in the real world.

In the past, humans were given instructions, commandments on how we should live, for example 'Thou shalt not kill' or 'Thou shalt not commit adultery'. Do we keep these laws rigorously? No, we do not.

But even though we humans are not perfect, the robot machines we originate or design and build are somehow supposed to be much nicer than we are, because that is essentially what the three laws of robotics imply.

For years and years humans have been building machines with which to make war. For example, when a heat-seeking missile is fired, it homes in on a source of heat such as an 'enemy' fighter plane, and closes in on the plane by tracking its engine heat. When it reaches the plane it explodes, thereby killing everybody on board. Is a heat-seeking missile a robot? If it is, then the missile has contravened law 1, so what do we do about it?

Perhaps a heat-seeking missile should have an in-built human detector on board such that before it blows up an enemy plane a megaphone appears from the missile and says, 'You are about to be blown to pieces, please eject from your plane.' Until the enemy pilot ejects, the missile cannot blow up the plane. Unfortunately the enemy pilot would not actually hear the missile asking him to eject, because his plane is travelling much faster than the speed of sound. However, if he did hear it he could simply stay on board, knowing that the missile will blow his plane up only if he does eject. Obviously this is a ridiculous situation, but it is an example of what must happen for Asimov's laws to hold.

But perhaps you feel that Asimov's laws only apply to robot manipulators, of the type on a production line. Many of these robots are extremely powerful devices, standing 6 ft high, and can swing around very quickly. They need, around them, safety cages and other safety features. This is necessary because if a human gets in the way of such a robot when it is in full swing, law number 1 can be quickly broken. In fact, when I was at Newcastle University in the early 1980s I became involved, as an expert witness, in a legal case on just this issue, following the death of a workman in a nearby company. The argument was all about who exactly was to blame for the death.

One issue that points to the impracticality of Asimov's laws in the real world is that of responsibility. Consider the case of a fairly intelligent robot owned by a human called Tyson. Tyson gets into a fight with another human called Bruno,★ and the robot watches as Tyson kills Bruno. The robot had the power to intervene but decided not to. In the eyes of the law in the UK and many other countries, the robot, if it had been human, would have been just as guilty as Tyson of killing Bruno. The robot remained silent, thereby consenting to the killing.

Tyson now teaches or programmes, whichever you like, the robot to protect him at all costs. To all intents and purposes the robot acts as a bodyguard. Tyson meets up with Lewis, and Lewis severely attacks Tyson, but the robot intervenes to such an extent that Lewis is killed, although this breaks Asimov's first rule. If the robot had not been able to stop Lewis from killing Tyson, it would have done a bad job. However, in doing its duty it also does a bad job. The robot is in a 'no-win' situation.

For a robot, or any machine for that matter, to move or take any action whatsoever there is the possibility that some human could be killed as a result.

We must be clear. Asimov's rules are fictional. There are no such laws which govern robots in the real world.

The issue of laws for robots does, though, raise the question of how intelligent robot machines will fit into our human world. I have already taken part in a couple of radio shows on just this issue, in South Africa and the USA. The main questions were: 'Should robot machines be able to decide on their own future?' and 'Should robot machines get a vote?' This topic was taken further in the debate at Robotix 96, in the Glasgow University Union in March 1996, on 'Will robot machines make better citizens in the next millennium?'

Let us take the last question first, though I will narrow down 'in the next millennium' to 'some time in the next century'. I take a 'citizen' to mean simply someone who lives in a collected

---

★ No link with any real-world people is assumed.

group such as a town or country. The definition can, however, also be expanded to encompass whether or not the individual has political rights.

The key follow-on question from the Union debate is 'Whose society will it be, humans' or machines'?' In other words, what will be the dominant life form? At present we have a human-dominated society, with human-driven values and human measures. Machines, therefore, play a subservient role. They are, at best, second-class citizens, they certainly do not yet have any voting rights and, for the most part, have no say in their own future. They do, however, help humans more and more to make their own decisions.

My own thesis is that at present we have a human-dominated society in which machines are subservient because humans are, overall, more intelligent than machines. Soon machines will, very likely, become more intelligent overall than humans, in which case we will then have a machine-dominated society in which humans, if they still exist, will be subservient.

In a machine-dominated society, machines and humans alike will be looked at in terms of machine-oriented values and machine-driven measures. A rough guess would say that some of these values will relate to human values. After all, humans originated machines. However, some of the values will probably not relate so easily because machines are, for the most part, different from humans. So the question of who will make the better citizens is a difficult one. Is it humans in a human-dominated society or machines in a machine-dominated society? The answer is probably that if a human asks the question, we will get one answer, whereas if a machine asks the same question, the answer will be different.

At the time of the Union debate Deep Blue, the computer chess-player, had just lost the first series of games with Gary Kasparov, having won their first encounter. One speaker suggested that Kasparov, being the better citizen, let Deep Blue win the first game in order to make things more exciting. My own reaction was that Deep Blue, being the even better citizen, let Kasparov win the overall series, having shown him in the first game who was really the better player. Obviously sporting events are open to interpretation.

In a human-dominated society, human rights and wrongs take pride of place. Human lives are important; lives of others, machines included, are much less so. In the machine-dominated society for which we are heading, we must expect that machine rights and wrongs will become the order of the day. We cannot, in reality as humans, expect anything more or better than this, although we may hope for it.

In a machine-dominated society, can humans expect political rights? This is extremely doubtful. Firstly, it is not necessarily the case that machines will have a political system as we know it today. With an overall far superior range of mathematical abilities, it is possible that machines will be able to look into all the different actions which can be taken in given circumstances and decide categorically what is the 'best' choice in terms of some previously defined function. Secondly, we, as much less intelligent creatures, would be unlikely to be given a say in how the far more intelligent machines run things. After all, the machines would be rapidly moving ahead, in terms of their overall intelligence.

But what about the short term? Are humans likely to give machines a political voice in the human-dominated world? Should machines be allowed to 'sit in' on committees which decide whether or not machines are given more of a say? It must be remembered that overall machines are, all the time, getting more powerful, more intelligent, are communicating with each other more and are themselves making more and more decisions about the lives of humans. Although we do not have to do anything in the immediate future because, to my knowledge, we do not yet have machines demanding suffrage, we may have to make such a decision in the years to come.

There is no guarantee that machines would want to vote in a human society, no matter how intelligent they become. However, if they do ask for a vote, one machine one vote, do we give it? I think that if we get to that stage, the answer has to be yes. It will be clear at that time that machines are a serious threat to the human world. Not to allow them to vote when they have the power and intellect to ask for it would, I have no doubt, incite a war that we would have no chance of winning in the long term. Or, as Hugo de Garris[59] feels, we may end up in a war between pro-machine and anti-machine humans.

We may, therefore, before too long, be faced with the choice of letting intelligent machines have more of a say in our human political future. Even now, many political decisions are based on the results of a computer machine's calculations. As pointed out in Chapter 6, many people could die or be put out of work because a computer system calculates a particular response. At present most computer systems of this type are simply programmed and merely work through a package of mathematics very quickly, completely under human control. But as these computers are allowed to learn, which is happening now, and to change their decisions based on what they have learned, they could very quickly end up with far more political power than our human elected representatives.

Perhaps in the next elections, whatever the country, as long as there is some form of democracy it would be sensible for us to know not only the policies of the candidates but also the machines and software that they will use. If a politician is subsequently to employ an artificial neural network then we should be told about it before an election.

Surely though, gradually handing over human political power to machines is just a continuation of a process that has been going on for decades or even centuries. In the eighteenth century, during what was called the Age of Enlightenment, the foundations were built for the profound effect of mass production in the twentieth century. Tasks were split up into a lot of smaller jobs. An overall production site operated much better with each human carrying out just one duty. This essentially is the basis for human society: you bake bread and I will grow vegetables.

All of this plays into the hands of machines, as machines can do specific tasks better than humans can. Now we have an Age of Machine Enlightenment. As humans, we split up our jobs and lives into small sections and gradually let machines take over these sections. Sometimes the excuse is that the job is dirty or hazardous and a human does not want to do it, but often a machine is simply far more efficient.

Sometimes, though, machinery can be of considerable benefit to humans and can improve the lives of some humans, for example people who have a disability. Where appropriate, a suitable machine can allow a person to do something that they otherwise could not do. Earlier in the book I mentioned projects such as the intelligent home and the intelligent wheelchair. These involve the use of technology of such a kind that benefit can be seen, without the worry or the side effect of machines making a further step forwards in taking over more of our lives in general.

But with this, how far do we let things go? A home has been described as a machine that you live in. An intelligent home then provides, eventually, a machine which can do all the thinking necessary in running a home. It can make all the necessary decisions to ensure that things go smoothly and can make sure that what is supposed to happen actually does happen. Is this something we want? Not only can a machine control heating and lighting automatically, it can also tell us that a caller is at the front door and even identify the person and decide whether or not to let that person in.

Homes around the world are now gradually being connected to cables, not only for television and telephone but also for the Internet. Many homes now contain a personal computer somewhere, although it is often mainly the younger generations which use it. It will not be long, therefore, before many of us have our homes connected into the network in terms of computer, television and telephone. Indeed, this has already happened for quite a number of people. Having such facilities at home reduces the need for office space and factory supervision. What can be done in an office that cannot be done from home?

Perhaps the one present advantage of offices, in which people come together to work, is human contact. Once that can be achieved in some way other than people actually, physically being together, then the advantages of large numbers of people not travelling to offices every day become more apparent. Virtual reality can provide us with the answer. If we can remain at home, but get enough of the interactive feeling that we are communicating with other people, then the need for offices just about disappears.

Offices cost money to heat and light. They take up space and are often in areas where there is no room for car parking or recreation.

Many factories are now largely automated, and are continuing to go in that direction, hence the need for humans at what was known as the workplace is reducing rapidly. In particular if, in the next few years, home-based offices become more widespread, then this will also save on fuel, will reduce pollution in the environment and will give (human) individuals much more leisure and/or family time. For people whose journey to work takes them an hour in each direction, the move to a home office environment gives them two hours extra every day for themselves. There are many immediate positives in humans working at home, not the least of which is that the companies concerned reduce the need for human-occupied buildings.

With virtual reality systems, research is presently going on to give individuals the chance, from their own base (possibly home) to come together virtually for a meeting, conference or discussion. Each individual can seem to touch the others, can see the others and each can respond appropriately to actions. Detailed information can be passed through computers, rather than on paper. The need to travel from Los Angeles to London for a meeting is removed.

This view of the immediate future achieves much which is positive for humans, but ties our dependence more heavily to machines. It is a continuation of what has been happening socially over the last 50 years or so in terms of the effect of technology, for example television, causing less everyday human interaction. However, it reverses the present trend towards more and more business travelling, without necessarily having any immediate effect on leisure travelling.

Once again, though, the result of such a change is to hand over more of the human role to machines. We will end up in our homes, sending messages to each other by computer, watching television and communicating with each other by telephone, videophone or a virtual reality system. Anything outside the home, such as communication or mobility requirements, will be done by machines, quite simply because they are better and more efficient than humans.

It appears that a natural result of all this is that some humans will be part of such a world much more than is the case now, and others will not. The others will be out of this technological world altogether; they will be second-class citizens. How will the humans who are part of the technological world deal with those who are not, and vice versa?

It will not be a problem for very long. It will be found that some of the tasks being carried out by humans in their homes are no longer needed, and in other cases the jobs can be better done by intelligent machines. After all, such machines will be able to make faster decisions, based on much more data and with much more precision and regularity. What we are heading towards is a machine-run world, in which all humans are eventually put aside, are not involved with what is going on, and in which all humans are, at best, second-class citizens. Who will be the first-class citizens? The machines, of course. After all, they will be more intelligent than we are.

Is there anything we can do to stem the tide? Is there anything we can do to prevent machines becoming more intelligent than we are? Certainly machines themselves will become more and more intelligent. There appear to be far too many driving forces, both financial and technological, to turn that around. But how about making ourselves more intelligent, or, as Moravec[25] has suggested, and which was considered in Chapter 8, shoe-horning ourselves into machine carcasses where our lives can continue in a new body and with a new brain?

Transferring our brain's arrangement, structure, synapse weightings and the like into a machine version would, in the first instance, produce two or more of a person. The original would, of course, lead a human life, presumably dying at a normal age. The version inside a machine would live as long as the machine lives. Indeed, if some way is developed of storing a plan of the brain arrangement in memory, then presumably that person, in terms of his or her brain set-up, could be stored away for several years and brought back to life in half a dozen or more different machines. All of this, though, requires the ability to obtain a detailed plan of the way a human brain is arranged. Technically we are at present a long way from achieving this. Indeed, we are only just about able to do such a thing with lower insects.

When we use a computer or a calculator, or even when we

write a book, we are effectively externalising certain features of our brain. We are memorising details in a concise form, and we know that we can retrieve these details, in a perfect state, at some time in the future. Conversely, we may wish to calculate some numbers or find the answer to a series of summations. Writing the numbers on paper or in a computer is a way of memorising them and displaying them more easily. Subsequently, when the sum is calculated, mathematical processing is carried out. As long as we had the ability and had been taught how to do these things, we could have done them inside our brain. Using a computer or pen and paper has merely given us extra, external brain power, in one form or another. However, the methods we have of communicating with such extra power are relatively very slow, by means of sight, touch and hearing.

In order to get a human brain's performance to increase, would it not be possible, therefore, to connect extra memory or extra processing capabilities directly on to the brain, possibly in the form of silicon chips? This is, I believe, a much more realistic suggestion than the Moravec idea. Although some knowledge of a brain's connectivity would be required, exact synapse weightings need not be known. More important would be knowledge of the general regional areas; that is, what each region of the brain does. If successful advances in machine technology were then complementary, any improvements could be taken in by the brain chips. Humans could in this way harness the thinking power of machines to work in their favour.

Implanting silicon chips directly into the brain, an idea popularised in the novels of William Gibson,[60] would not seem so sensible. The size of a human brain has evolved to be about right for what it does, so we should consider clip-on chips to the brain rather than implants. Different chips could then be used for different things, one for memory, one for processing skills, one for musical abilities and so on. Perhaps various sockets could be placed over recipients' heads and they could plug in different chips dependent on what they required. They would not need to watch a film, they could simply plug in a relevant memory chip and they would instantly have seen it.

A big advantage of brain chips would be speed. Instead of

taking a relatively long time, as at present, to get from a thought, through our fingertips to operate computer keys, and then to see a solution on a monitor, it could all be done internally with the brain chips as part of the internal process. Unfortunately some of the original brain power would necessarily be used up in learning how to interact with the brain chips. However, if the brain chips themselves have significantly increased powers, the overall effect would be very positive and would certainly give a person extra brain capabilities.

Sockets on a human brain could also be used for receiving different signals, such as X-ray, radio or radar, as was suggested in Chapter 10. Humans could then communicate with each other by sending brain signals directly, by radio waves or even infrared. In this way they would become more like the Seven Dwarf robots. Certainly we would not need transistor radios any more, as a person could mentally tune in directly to the radio station of their choice.

But surely all of this is far-fetched? Frankly, yes it is, but less so than completely transferring human brain arrangements to machines. In order to become effective, it requires major steps forward not only in understanding how the human brain is arranged on the inside but also how to plug into it successfully in a way which would help the brain's overall operation.

It is known that some fish produce electric fields to communicate. Sharks home in on prey in this way, and birds such as pigeons, and even bacteria, can navigate by means of the earth's magnetic field. Would it not be good if humans could plug in extra brain chips that allowed us to use such senses? But a major concern is that this would imbalance the present human brain set-up. Taking in other signals and processing or understanding them would require some of our present neurons to be redeployed. The same would be true of brain chips with extra memory or extra processing skills. In each case our brain operation in several areas of present use would have to decrease in order that we could accommodate new skills. The new inclusions would effectively be special-purpose partitioned brain blocks, adding to the total brain size but taking up present neurons in order to be operative. We would have to lose something in order to gain something else.

So, realistically there appears to be no human comeback to the rapid increase in the size and power of intelligent machines. On the assumption that machines will be, after some time, more intelligent than humans, then we have no apparent answer. Humans are severely restricted by our biological framework. We simply cannot, it appears, grow human brains with more neurons, which can operate in some ways better than they do now. Even if we do, they will, before long, be outperformed by machines. Perhaps we could increase the performance of human brains by highly selective training, maybe using virtual reality to push the training to the limits. But importantly there are limits, and such an action is simply buying us humans time.

In Chapter 5 we looked at consciousness and, in particular, human consciousness. A conclusion appears to have been drawn by some[34] that because human consciousness cannot be *exactly* copied by a machine, so machines will always be subservient to humans. On the assumption that dogs are conscious in a doglike way and we humans cannot *exactly* copy a dog's consciousness, it appears also to follow that humans will always be subservient to dogs. Obviously such a conclusion would be ridiculous, just as is the first conclusion. No, machines will not always be subservient to humans because they are not conscious exactly like humans. Indeed, a cruise missile is only subservient to the humans firing it, and is not at all subservient to the target, whether human or otherwise, unless the target is even more intelligent and knows how to control the missile.

A key issue is who is in control, and between species it appears to be that the more dominant species tend to be more intelligent. Indeed, the same is true within a species. Each human is conscious in a humanlike way, each dog in a doglike way and each bee in a beelike way. When switched on each machine is, if you like, conscious in a machinelike way. But consciousness is fairly abstract and is not really something we can easily measure. It is not, however, something we should hide behind. Humans do not control dogs because we are more conscious than they are, we do so because usually humans are more intelligent and possibly have some physical advantages.

Intelligence is something that can be measured, and we have

been doing so for many years, often badly. Unfortunately, in taking such measurements it appears that on any single unbiased measure, machines can outperform humans. The one thing humans have at present, however, is that we can perform well on a very wide range of tests. An important point appears to be, therefore, that overall, humans can perform with some degree of intelligence over a wide variety of tasks. Generally, intelligence (IQ) tests look simply at only one, two or a few tasks. Indeed, the same is true for all academic examinations. As I have said before, by doing well in a test a human shows he can perform that task well, but that is all he shows.

I am sure you have heard the expression 'absent-minded professor', meaning a professor who thinks so deeply about his own subject that he is vague about other things. Possibly he cannot remember his partner's birthday, or he crashes his car because he is thinking of something else. Yet that person has shown that he excels in one, or a few, subject areas. The same is true perhaps of doctors and other professions. Being a professor or doctor simply means that that person can do well in his profession; it does not mean that he is wise on all, or even a wide range of, subjects. But as he is human, he can perform reasonably on a wide variety of things.

The person with autism portrayed in *Rain Man* performed very well on certain things, but was unable to perform reasonably on others. It was this latter inability that cast the character as mentally disabled. Professors, doctors, the character in *Rain Man* and many other people with a mental disability have common ground in that they can excel in certain disciplines. It was only his poor performance on other things that let the *Rain Man* character down, the same result being true for many people who are mentally disabled. A person who is not disabled can do lots of things reasonably well. A mentally disabled person cannot do some of these things or can only do them poorly, even though he may be able to do other things very well. Comparing humans with machines, we have noted that most machines are presently designed to do one, or only a few, things excellently. If, however, we consider whether or not a machine can be more intelligent than a human, we have to look at performance over a range of tasks.

With an overall intelligent machine, therefore, we are looking at a machine that can do a variety of tasks in an intelligent way. If we have a machine that does just one intelligent thing very well, as long as that one thing is not too injurious, a human can probably control it by performing better in many other ways.

A conclusion to be drawn from this is that all of the elements of a machine which are far more intelligent than any human are ready and waiting. They have simply not yet been put together. Why not? Because we do not as yet know how.

Let us take a large artificial neural network which deals with the vision system of a robot, another that moves the robot around and finally one that aims and fires a machine-gun. Can such things be done? Yes, of course they can. Will the robot work? Yes, after a fashion, and we would probably get coordinated action by linking the neural networks together. But there are 1,001 ways in which such a machine could be stopped, particularly if it could not hear us or did not recognise a bomb. How the overall brain is connected together is critical, just as is the range of senses feeding into that brain which such a robot would need.

In the film *Terminator*, Arnold Schwarzenegger plays the part of a Cyborg, part machine and part human. He is mainly a machine, although conveniently it is his outside which is human, with flesh, skin, hair and blood. His inside, including his brain, is machine, with full armouring, and is therefore very tough! Leaving aside some issues in the film, such as time travel, the question could be asked, 'In reality, could a Terminator be produced now?'

Firstly, it would not be necessary for a real Terminator to be humanlike to any great extent. Certainly it would not need to look like Arnold Schwarzenegger. It would not need a biological exterior with skin, hair and so on. In fact, it would not need a human form and shape at all. There is no reason why it could not have six legs, rather than two, and wheels as well for that matter. It could be a completely different size from a human and would not need to be armoured, as it could use other machines for both protection and weaponry.

What would a real Terminator need? Initially it would have to have a general level of intelligence at least on a par with most if not all humans. Perhaps it could be more intelligent. It would also need to have a fairly broad range of sensory abilities, some humanlike, but some extras would be advantageous. Overall it would not need to be physically superior to humans. It could actually be inferior in a number of ways. After all, because of its intelligence it could work out ways of defeating superior fire power.

Importantly there would need to be more than one machine, which would not necessarily be the same. This is a big advantage for machines in that they can be very different, both physically and mentally. Presumably the machines would be able to communicate with each other, not only in a way that humans could not understand, but also much more rapidly and efficiently than humans do (without the aid of machines).

So our real Terminator machines are not necessarily much like a human. It is said in the film, 'The Terminator is out there. It can't be bargained with, can't be reasoned with, it doesn't feel pity or remorse or fear and it absolutely will not stop, ever, until you are dead.' In the film, the Terminator is simply looking for one, and only one, person, whom it must terminate. It doesn't even know what the person looks like, all it has to go on is a name, and as a result it terminates everyone with that name. But how does this relate to our real Terminator?

Put yourself in the position of being a very intelligent machine. You are part of a new breed. A group called humans still exists and there are many of them. Having been the dominant life form on Earth for many years, they are not too keen to give in to the new breed, even though they originated it. They are trying as hard as they can, therefore, to destroy every single member of the new breed. Their idea is that if they can destroy them all, then humans will again be the dominant life form, and next time they won't make a hash of it!

So what do you and the other members of the new breed do? You could try to be nice to the humans. Maybe in time they will be happy to be second-class citizens. But why should you? The humans are less intelligent than you are, and given half a chance they would most likely try and end your life. It is quite dangerous to give humans any power at all, as they would be likely to use it against the new breed. Some boisterous humans will try to do something against you anyway, and what should be done with them? Should you show fear and run away, show remorse, discuss the pros and cons of such an action, or should you take action yourself?

In reality we can only speculate as to how members of the new breed would treat humans. After all, the new breed is more intelligent than we are, so it is difficult to judge. Perhaps the most fruitful approach is to look at what has happened in the past and to project this forward. Nietzsche[61] gave an apt description when he said firstly, 'All creatures hitherto have created something beyond themselves.' He went on, 'What is the ape to man? A laughing-stock or a painful embarrassment? And just so shall man be to the superman: a laughing-stock or a painful embarrassment.' The superman in Nietzsche's words is likened here to the new breed of machine.

So, to get a best-guess picture of how the new breed machines would treat humans we should look at how humans have treated those from whom we have evolved. What do we do with chimpanzees or other animals? Do we treat them like brothers? Do we regard them as our equal and elect them to government? Apart from one or two exceptions we certainly do not, but why should we? They are less intelligent than humans. It would be an embarrassment to have an orang-utan as prime minister or president. What we do to apes and our other distant evolutionary relatives is shoot them, remove their natural living environment, cage them and ourselves stand outside the cage and glare at them. We use other mammals to make our lives more pleasant, killing them for food and controlling their lives in every respect. In the UK, foxes are hunted and killed, even now, just for sport.

Most likely the apes have been nothing like the same threat to humans as humans will be to new breed machines. Gangs of apes have not, in general, gone around hunting down humans in order to kill them. Despite this, humans have gone out hunting to find the apes and have killed or caught them, even though they have not been able to fight back with any strength or power which compares with that of humans. So if humans decide to fight the new breed machines, what would we expect? If you were a new breed machine would you trust a human? Would you allow less intelligent humans to tell you what to do, or, as you know better

273

than they do, would you wish to tell them what to do? We must expect that the machines would probably wish to dominate and that they would force home their domination in both mental and physical ways. This is the way we humans behave at present, and as the machines are more intelligent, surely they would learn from our experience.

It was only in 1859 that Charles Darwin[62] published his famous *Origin of Species by Means of Natural Selection*, in which it was proposed that new species arose by a process of natural selection acting on individual inheritable variations. It has later been shown that the changes probably result from spontaneous genetic mutations. Evolution as a whole then consists of gradual changes in species along with the introduction of new species through mutations. As a result of this the diversity of plant and animal life on the Earth at present, humans included, has evolved from the first primitive living organisms. Evolution, and in particular Darwin's contribution to our understanding of it, is now widely accepted. Changes in species and even new species have occurred in the past and will occur in the future.

But evolution does not necessarily stop with humans as the dominant life form; indeed, it surely will not. Just as dinosaurs came, were powerful and subsequently went, so too will humans have their day. Perhaps we will simply die out for one reason or another, but it is more likely that we will be superseded. A new breed will appear, that is a superior species to humans, whose members are more intelligent than humans. Now, as at no time in the past, humans have considerable control over biological mutations, but the likelihood of a sustainable biological mutation occurring from humans is, I feel, extremely low. We could, however, cause the transition to machines; we could easily hand over control to machines. Evolution works in similar ways in machine design as it does in a biological sense, except that generations can pass in a very short time span, as was explained in the last chapter. With machine evolution, the population can be strictly controlled and new machine designs can result from directed evolution. Designer evolution is the order of the day for machines.

So the evolutionary theory of Darwin and the philosophy of

Nietzsche between them point to the fact that humans will be superseded as the dominant life form on Earth and that whatever becomes the dominant life form will not treat humans with much respect. It is because of our intelligence that we humans are the dominant life form on Earth, and should there arise a more intelligent life form then the chances are that it will dominate humans, in the same way that humans dominate less intelligent species. There appears to be nothing to stop machines becoming more intelligent than humans in the not-too-distant future. So what can we conclude other than that machines will dominate the Earth? Not only this, but it will not be very long before they do so.

But machines are a very broad selection of different things, ranging from very simple devices to the more intelligent Terminator type of robots which have been considered here. Indeed, this range of types is a distinct advantage for machines. As humans we have one body form, around which we have had to design machines so that we can operate widely in the outside world. Essentially we use machines as extensions of our human capabilities. As a result, we have given over more control of our environments to machines and enabled them to move and operate effectively within these environments.

It would be wrong to think of robot machines solely in terms of science-fiction-type beings, which look as they do either because they are merely humans in disguise, as with the Terminator or C3PO in *Star Wars*, or because practical limitations on film-makers cause them to look like dustbins, as with R2D2 in *Star Wars* or Dr Who's famous enemies, the Daleks. In most cases, however, the fictional robots' physical and mental abilities have been modelled on those of a human.

But to obtain a realistic view of the future, we must consider not only the wide range of intelligent machines already in existence, but also the whole gamut of possible developments. We know that there exist very large, very powerful computational-type machines. We know that there also exists large-scale communication networking. Finally we know that there exists a wide variety of robotic machines of different shapes and sizes. It is the combination of these different elements which is the threat to humankind. At present we are in control. However, as soon as control of any one of these three sectors starts to get tenuous, our real problems will begin.

The robot machines which have a physical, moving presence do not in themselves need to be humanlike. They do not need to be self-sufficient in the way humans are, or even to be particularly intelligent. What is important is their overall intelligence capabilities and the overall self-sufficiency of a machine-controlled network taken together. Machines can have their intelligence distributed in a way that humans cannot. Even Walter and Elma, the walking robots, have legs with a little bit of intelligence in them. In general, because of the superior communicating abilities of machines, intelligence can be spread around a network. Local intelligence may deal with local issues but will have ready access to more powerful capabilities.

Humans have in fact been using computing machines in this way for quite some time. Historically, one central powerful mainframe computer was hooked into by a number of fairly dumb monitors. To all intents and purposes, virtually all the processing was done in the mainframe. Now, however, the trend is for computer workstations or personal computers to be positioned where required, sometimes in clusters, each machine containing fairly powerful capabilities of its own. Each workstation or cluster is then connected up to a central machine which is used chiefly for communicating with other machines, although they can also tap into some fairly specific processing. Importantly, though, these are still under human control. However, the network is in place.

With distributed machine intelligence, the overall capabilities are partitioned, with each element or subsection contributing to the total. There may be physical similarities between the elements, as in the case of computer workstations. However, if robot machines in general are also linked into the network, the physical spread, the range of intelligence levels and the duties performed can be very broad. Each element certainly adds to the total, but not in terms of doing a similar sort of thing in a very similar sort of body, as in the case of bees in a hive, but rather in terms of carrying out a particular function in an appropriate physical form.

All this means that robot machines, of the Terminator type, would not in reality need extremely high levels of on-board intelligence, but could link in with and communicate with further levels of intelligence via a network. Only locally necessary intelligence capabilities would be needed on board. Further, the robot machines could be of different shapes and sizes. At the lower end, perhaps, are micro-machines, which are presently being designed to move around inside a human body and, for example, operate on nerve cells or cells in the retina of a human eye. At the higher end, perhaps, are machines such as aircraft, missiles, tanks and rockets. The important point is not their individual, on-board intelligence, but the overall intelligence of machine and controller together, whether the controller be human or machine.

When, therefore, we consider whether a machine is more intelligent than a human, we do not necessarily look at an individual computer sitting on a table, but rather at a network which could contain many computers and many robot machines. Further, when we consider intelligent machines evolving from humans and becoming the dominant life form on Earth, it is not so much in terms of a troop of humanoid Terminator robots, but rather a network of machines with computational and other physical abilities, including robot machines of various shapes and sizes. Intelligence is spread around the network. The only way to beat the system is to destroy the network as a whole. Taking out one robot will do very little good.

But for such a machine network to take over in the first place, all three elements appear to be necessary. It is difficult to see how a troop of Terminator robots alone could become dominant, as humans would have networking, communications and computers under their control. Even if each robot machine was more intelligent than a human, they would surely need to communicate in some way in order to dominate as a species. United they stand, divided they fall. Once connected via a network, it would be sensible for robot machines to harness processing power on the network, if only for the same reasons as humans.

However, it would be difficult to see networked computing power, with no physical robot machine elements, taking over on its own, unless it relied on human slaves carrying out its will. At times, I wonder if we are not far from that situation now. However, humans still appear to have the edge in terms of intelligence, that we are, in a way, slaves through choice, and that someone can still switch off the network and walk away from it.

For complete machine domination, therefore, it appears that all the elements must come together. That is, computing power needs to network with robot machines, the overall total having distributed intelligence capabilities which are superior to those of humans, even when humans are networked into other machinery under human control. But these elements *are* coming together. We humans are bringing them together, for our own gain. All that remains is for the intelligence of the machines to become greater than that of humans.

So we are not looking at a Sorcerer's Apprentice scenario, or where a mad scientist in a laboratory builds a Frankenstein-type machine which shoots everybody and then runs the world. We are looking at a variety of machines which, rather like bees or ants, link together for an overall cause. Just like bees or ants, each machine could have a relatively low intelligence, but unlike them, each machine could be very different. It would be difficult to reason with or argue with a land mine which is networked, which is linked to a higher-level system that causes the mine to explode when only a human is near. Just so, it is difficult to reason with a cruise missile that is heading straight towards you. But then, these are both machines, machines of destruction. For any human it is important who is controlling such a machine: the 'home team', another group of humans or an intelligent machine network.

We are not looking towards machines which tend to be human copies both physically and mentally. In order to appreciate our situation we need to move away from the thought that as we are human we are the best there is ever going to be and that nothing will evolve from us. There is no mysterious process going on in a human brain which causes us to be conscious. A human brain is simply a physical entity. We are what we are: a state of brain and body activity. We are simply part of the evolutionary chain.

Apart from an overall level of intelligence, where humans still seem to have the upper hand, much of what has been talked about in this chapter is science fact, not science fiction. Whether we like it or not, we have land mines, an estimated 200 million in 60 countries, although we cannot find most of them.[63] Whether we like it or not, we have missiles. We also have robots which can immediately pass on what they learn to other robots on the other side of the world. This is all technology which is with us now. We also have networking around the world and large, powerful computers connected to it, making decisions which will affect the future. Are we really far from the time when machines will dominate?

Intelligence is the key. At present humans have the edge and are therefore in the driving seat. But once machine intelligence is comparable to that of humans, then the human race will come out of the Indian summer it is presently enjoying and will find itself in a never-ending winter. To all intents and purposes, the human race, as we know it, will be no more.

Once a comparable level of intelligence is reached, machines have the capabilities to go way beyond the achievements of humans, both physically and mentally. Even the frontiers of space can be seen in a different way when the frailties and limitations of humans are removed. And for all of this we are not looking a million or even a thousand years ahead, but just into the next century, maybe in the next 25 years, and very likely before 2050.

So the human race will probably, in the lifetime of many of us, be surpassed by a network of intelligent machines, which we originated. Many things which we humans do at present are not only aiding but are even speeding up the process; for example, the design of massively parallel computers, more powerful weaponry and even micro-machines. But are we humans not intelligent enough to realise what we are doing, where we are taking things? Surely we cannot be stupid enough to orchestrate our own destruction?

Chapter Thirteen
Mankind's Last Stand?

The main points of this book can be summarised as follows:

1. We humans are presently the dominant life form on Earth because of our overall intelligence.
2. It is possible for machines to become more intelligent than humans in the reasonably near future.
3. Machines will then become the dominant life form on Earth.

If asked what possibility we have of avoiding point 3, perhaps I can misquote boxing promoter Don King[64] by saying that humans have two chances, slim and a dog's, and Slim is out of town.

Point 3 and the Don King conclusion are supported on point 2 by the observation that a human brain does not have some special magical chamber; it simply exists as a physical entity. A human's conscious awareness is essentially the overriding mental state in his or her brain. Given sufficient processing capabilities (neurons), overall brain size and complexity of connections there is, therefore, no reason why a machine or, for that matter, another biological being cannot be just as intelligent as, or far more intelligent than, a human. It is accepted, though, that machines may need a more detailed low-level model of the brain's elements than a neuron. However, as discussed in Chapter 5, my own feelings are that neurons should be good enough, in that it is not required to copy a human brain but rather that a machine brain should be more intelligent. At present we already have good neuron models. Within the next ten years there will be machines with a brain equivalent in size to a human's, in terms of quantity and density of neurons. So brain complexity, how a brain is put together, is the one remaining issue.

Supporting evidence for this has been given by looking at some of the similarities between human and machine brain operation. Certain humans – for example, the *Rain Man* character – function in what we have thought of as a machinelike way. However, it is quite possible for machines to carry out many functions, although at present more in a simple animal or insectlike way than in a humanlike way.

The Reading robots have been specifically aimed at showing basic brain functions operating together, especially learning and communicating. The robots have exhibited a range of behaviours which can be directly likened to those of insects and mammals, including humans. Group and evolutionary behaviours are extremely prominent. These robots have been used to prove what is possible, nothing more and nothing less.

The Reading robots are cute, friendly and popular, but these characteristics are not essential. Even now a not-so-nice group of researchers at another university or military base could possibly build robot machines with the same behaviours, group characteristics and leadership qualities, but on a much bigger scale and with weaponry attached. There *is*, therefore, something to worry about. The same is true if intelligence is built into weapon systems such as missiles, fighter planes or tanks – and this has been going on for quite some time.

Communications between machines have been pointed out as being much better than those between humans, unless humans also use machines. Machines can communicate much more rapidly and use less energy. Machines can communicate in many different media, some of which, such as infrared or ultrasonic signals, are beyond human capabilities. All communications can, however, suffer from disturbances of one form or another. In a crowded room it can be difficult for one human to hear what another is saying, because of other people talking or a machine playing music. In the same way, infrared signalling can be disturbed by bright light. In both cases, if a message is not received well it may either be misinterpreted or the intended recipient can ask that the message be repeated. In the machine's case this can occur very quickly.

In the past, humans developed Morse code, which was used for telegraphy, realising a benefit to humans. Later, telephones provided a greater benefit, as have radio, television, satellite communication systems, and so on. In each case, machines combined with physical properties of the world have been employed under human control for the human good. So surely, in the case of intelligent machines, these can also be made use of for human benefit!

Up to a point intelligent machines can be, and indeed are being, used for the human good, particularly as aids for people with disabilities and for intelligent homes. But this is all on condition that the intelligent machine is not more intelligent than the human controlling it. As soon as the machines become, overall, more intelligent, then we have a different situation. Once some machines start to be more intelligent, overall, than some humans, things will begin to become difficult. We may not be far from that point now.

This all raises the question of who is in control, who does the switching on or off? We may get to the stage of a two-class society: those humans who can control intelligent machines and those who cannot. However, in the last decade or so we have been moving towards a dependency on machines which are never switched off and even have back-up power supplies in case problems occur. So perhaps humans will design the off switch out of machines before the machines design out the switch themselves. Certainly a really intelligent machine will itself figure out some way to avoid being switched off, if that is important to it! .

Some people worry about the whole issue of machines themselves designing, building and operating future machines. Unfortunately there does not appear to be any way to avoid such an eventuality. Indeed, to an extent this happens now, although at present humans appear to be in control at various points along the line. We have, though, numerous cases in which a human merely starts off a process by, say, selecting a colour or size. A button is pressed and shortly afterwards a new machine, such as a car, appears, having been designed, constructed and tested completely by machines. It is only one small step for an intelligent machine to make the decision to *start*, with an outcome that is not a car but another intelligent machine of some kind.

Apart from the concept of several machines coming together so that good features from each of them can be grouped together like genes to produce a new, super-machine, the control of machines building machines is critical. If the machines doing the making are strictly controlled, as in most cases at present, with directly programmed instructions, then there is not really a problem. If, however, these machines are intelligent to the extent that they can, under their own control, change the design dependent on their view of external results fed back to them, possibly by other machines, then machines can build better machines. The better machines can then either themselves build even better machines or can simply ensure that better machines are made.

So not only can machines build better machines but the initial, perhaps human-originated machines can design and build their own successors, which are more accurate, faster and even more intelligent. The successors will then build even better, and even more intelligent further successors, and so on. Although it appears to be extremely important for humans to avoid getting into the situation described, the clear, present financial drive is on for exactly that to happen. Humans cannot now design and build machines as well as other machines can. It is better, therefore, indeed it is progress, to get the machines to do it all themselves. Once again it points to machines far outperforming humans before very long.

It was previously mentioned that machines can use less energy for communication. It is quite possible, also, for successor machines to evolve generally to be energy-efficient and eco-logically friendly. This could easily provide a strong ecological justification for machine supremacy. Quite simply, if machines can operate using less energy than humans, then they might want to take over in order to save the planet.

One question asked is: 'Is there any reason to believe that non-organic life [machines] will develop a will to survive and compete analogous to that in organic life, or is it just that machines are inevitably taking on our frame of reference?'[65] This is an important question because although certain machines may be of a similar level of intelligence to a human, this does not immediately imply that they have a will to survive. However, we humans, other mammals and insects appear to have such a will. Why do you wish to stay alive when you wake up in the morning?

I feel that humans have such a survival feature as a basic instinct passed through genes, from long ago. Any creatures, including humans, which did not have such a strong will would have died off quickly, without reproducing and, therefore, without passing on their 'not-too-bothered-about-life' or even 'death-wish' genes. So it is the 'will-to-survive' genes that have been passed on from generation to generation.

It is true that some humans would initially set up the gene program for a breed of intelligent machines, and could seed the machines perhaps with 'not-bothered-with-life' genes. However, after a while, if one of the machines had modified its behaviour slightly so that it became at least a little bit peeved at being attacked, then it would be much more likely to survive than a machine whose behaviour had veered slightly in the direction of cutting its own wires.

Whether or not, therefore, a human frame of reference is passed on to a machine, the survival of the fittest quickly comes into play for both organic and non-organic life. The survival instincts or behaviours are more likely to be passed on by the fittest. It is more likely the fittest machines that will be successful in making further machines. The human frame of reference is an important practical aspect, though, in that we are unlikely to put together machines with suicidal or self-destruction tendencies as it does not make immediate economic sense, even if they are merely given, for example, two years of normal life and then an instinct to commit suicide.

In Čapek's play, *Rossum's Universal Robots*, the robots get to a stage where they are about to take over the world. However, they stop and ask the question, 'What are we going to do next?' The robots realise that they do not know what to do with themselves and, therefore, still need humans in order to give them a reason for being. This conclusion, which gives the story a happy ending (for humans), relies on the fact that humans themselves have a reason for living, a self-will, but the robots do not. From the discussion, it follows that machine evolutionary improvements would, quite straightforwardly, give Rossum's robots the will to survive and dominate, thereby suggesting that the play's ending could, perhaps, be rewritten, or that Chapter 2 of this book could act as a sequel.

There are still, I feel, some humans who believe that even if Rossum's robots did exist as in the story, they would not take over things because we humans could simply switch them off, just as we would a computer workstation on a table. If you are such a person, then please think again. If the computer workstation had arms, legs, sensors and a machine-gun, then would it let you switch it off if it did not want you to, or even if it had simply been programmed to prevent you switching it off? As another example, if a large missile was hurtling towards you, how would you switch that off? Perhaps you would not even know it was coming until it actually arrived!

Missiles and other military weapons and vehicles are obviously an important driving force in the push for more and more intelligent machines. It is also an area in which one can quite simply realise what the effects of more intelligence in machines could be, much more so than in the case of, for example, large computers which operate on the international stock market. In the Gulf War's Operation Desert Storm, which started in January 1991, a wide range of weapons and vehicles was actually employed, in which computer-based systems were used to make plans or decisions, to adjust target tracking and to home in on identified enemy machinery. One example is the US's McDonnell Douglas AGM-84 SLAM,[66] an air-to-surface missile which was successfully deployed against a number of Iraqi targets during the war. The turbojet missile uses the Global Positioning System (GPS), so that *it* knows exactly where it is, a data link on which to pass and receive information, and an infrared imaging (IIR) homing device with which it can inspect an infrared picture of its target. With the IIR it initially selects its target, and then homes in automatically. The aircraft from which the missile is launched needs to get no closer than a few hundred kilometres from the target.

The SLAM is just one example, however, of presently available missiles. The Hughes AGM-129 carries a 200 kt (kiloton) nuclear warhead, can be launched 3,000 km from its target and also has laser radar tracking. Inbuilt computer-based tracking systems are also housed on France's ASMP, a 300 kt nuclear missile, and Russia's Raduga KSR-5N, a 350 kt nuclear missile. Perhaps one of the most famous missiles is the Tomahawk BGM-109G cruise missile, which has a range of 2,500 km (1,550 miles) and flies at a speed of 550 m.p.h. The most recent versions of this mainly ground-launched missile have thermonuclear warheads and match the target scene with a picture already held. In this way they can achieve extremely high accuracy. Back in July 1984, in a demonstration, a sea-launched Tomahawk was shown to impact precisely with a small concrete wall (its land target), after a 400-mile flight. On an even bigger scale, however, Martin Marietta's Titan II Inter-Continental Ballistic Missile,[67] now old technology, carried a 9mt (megaton) warhead and could travel for 9,300 miles using a space guidance system. It is extremely difficult to see how an individual could 'switch off' a Titan II missile when it was flying towards him at 15,000 m.p.h.!

But it is not just missiles in which a considerable amount of computing is used for military purposes. The M1 Abrams main battle tank houses a laser rangefinder for its main armament, a 105 mm gun. When a shell is fired from the gun, adjustments and targeting are all carried out by a computer in order to ensure that the target is hit. As far as military machines are concerned, most of us only get to find out about a small fraction of the things that are actually going on, and we get merely a glimpse of the weaponry available. However, based on our knowledge of what is, and what is not, technically possible, we can all hazard a guess at the possibilities, particularly with respect to the military use of intelligent systems.

Perhaps the greatest amount of technology employed within the military domain is that in combat aircraft.[68] The flight control of most present-day warplanes is based on what is called 'relaxed static stability'. Examples of such machines are the Eurofighter

EF-2000, the F-22 Rapier, the Russian Sukhoi Su-27 and the US General Dynamics F-16 Fighting Falcon. Relaxed static stability means that if the plane was not controlled it would quite simply crash, so it must be flown actively. However, in all of these planes, active flying is well beyond the means of a human pilot, so fly-by-wire or fly-by-fibre optic systems are employed for controllability and agility. As a result of this and other computer control systems, the pilot's housekeeping duties, to keep the plane in the air, have now been largely taken over by the on-board computing. The only information actually given out by the system flying the plane is that indicating that a condition value has strayed out of its normal operating range such that some corrective action may be necessary.

In modern-day warplanes the pilot's role is mainly to ensure that the tactical aspects of his mission are carried out appropriately. Flying, targeting and firing are all largely carried out by computer. Essentially the technology involved has moved on so far and the speed and complexity of operation of the modern warplane are so great, that all a pilot can really cope with is the tactical side of things. Machines are better at the remainder of the problem!

This raises the question as to why combat pilots still exist. Many missiles have data links which send data to and receive data from a ground station. They can, therefore, be controlled remotely or, as is now the case, once fired organise themselves through intelligent computer control. Small spy planes, rather akin to enlarged model aircraft, are also flown remotely and are designed to be pilotless, so why should each combat aircraft need a pilot?

In fact, combat aircraft presently under development[66] are designed to have no pilot on board. One example is the McDonnell Douglas/NASA X-36 which uses thrust vectoring for flight control and is designed to fly without a pilot at a fairly low speed but to have a high angle of attack. It is aimed at being difficult to pick up on radar, in common with the Lockheed F-117A Nighthawk (Stealth) fighter, but has a long operating range and is highly manoeuvrable. Importantly, without a pilot *in situ*, many safety features are no longer required and the plane is not limited by the relatively poor performance of humans, particularly in terms of speed of reaction.

Although, by going pilotless, it appears that the differences between combat aircraft and missiles are being reduced, fighter aircraft are extremely versatile, with a wide range of capabilities. For instance, they can go into combat against a manoeuvrable enemy such as an opposing fighter aircraft. Combat aircraft also have a range of fire power at their disposable. For example, each F-16 houses the following: one 20 mm Vulcan cannon, nuclear weapons, air-to-surface missiles, air-to-air missiles, anti-radar missiles, anti-ship missiles, free-fall or guided bombs, cluster bombs, dispenser weapons, rocket launchers and napalm tanks.[68] It is frightening (for humans) to consider such a collection of weaponry on a piloted enemy aircraft. However, it is clearly the case that before long, intelligent machines, most likely computer-based, will not only be flying such combat aircraft, but also making the decisions as to when to fire the nuclear weapons and at which target, human or otherwise, to fire them.

With data available to be communicated to and from a combat aircraft, it is, of course, not necessary that the intelligence to operate the aircraft is actually carried on the plane itself. It is quite possible for the aircraft to be controlled remotely, using a link into a data network through which decisions can be made and commands given. Interestingly, this is getting very close to the type of set-up I discussed in the previous chapter, in terms of the overall requirements for intelligent machines actually to take over from humans. It is worth pointing out, though, that all I have described here is generally known present-day military machinery and generally known present-day technology.

With pilotless combat aircraft, missiles and quite possibly unmanned tanks and other military machines, we face an interesting near-future war scenario. Assuming one group of humans wages war on another group of humans, perhaps in the form of two groups of allied international powers, then the whole war picture will be remarkably different from that in the past. We will see intelligent self-controlled aircraft fighting against other intelligent self-controlled aircraft. We could see autonomous intelligent tanks fighting against other autonomous intelligent tanks. Certainly military strategy will play an important role, but it will be intelligent computer-based military strategy, with little or no human intervention. The whole war picture will be more like a chess game, with one military version, of Deep Blue maybe, fighting against an opposing military computer-based system. Instead of pawns, though, we have autonomous tanks. Instead of bishops we have missiles and to replace rooks we have pilotless combat aircraft. Certainly strategy will play a role, but it will be a computer-based strategy because that is the best available. The computer-based technology itself will be important: how quickly will it operate, how accurate will it be? The original program will be a key component, but so too will be what the system has been able to learn, perhaps even how it has been trained.

So, in wars to come, technology will fight technology. In combat, the aircraft which is technologically best, in terms of fire power and manoeuvrability but chiefly in terms of intelligent behaviour, will win. And once one side has forced its way through, once it has shown its dominance, the opposition will be at its mercy. If the technology thrown at the dominant side has failed, the defeated humans will be able to offer little or no further resistance. They will be dependent on the fact that the dominant side is human-controlled, that the humans have overriding command over their machinery and that these same humans are, after all, not so bad. One should remember, however, that it is exactly this overriding command that is presently being designed *out* of military machines.

In Operation Desert Storm, if reports are to be believed, one side appeared to have a clear technological edge over the other. This led to one side not only having supremacy in the air but also gaining complete control of the war as a whole, in a way not previously witnessed. We can be sure that the capabilities of military machinery have improved considerably since that time, some of which the general public knows about and most of which it does not.

But as long as humans do have overriding control, even if a future war is merely a case of 'our machines versus their machines', we will still be all right in the long run because both sides can stop their machines when they want to. However, can we rely on this? If their machines are better than ours, if their machines have more fire power, if they can act in a much more intelligent way, then what can we do about it? Surely our aim is to get one step ahead, to put more fire power into our own machines and ensure that they are more intelligent than the enemy's. This has always been so; indeed, it is an important feature of military conflict.

In the days of the Cold War, agreements were made between the major powers to limit the number and extent of weaponry, particularly nuclear missiles, in order to try and put a cap on this military escalation. But do we presently limit the amount of intelligence put into machinery? No, we do not. How can we, when we do not really know how to measure it and we cannot really tell when one person or one machine is more intelligent than another? So, without any form of checking or control, more and more intelligence and autonomy will be given to military machinery of all kinds, simply to stay ahead of the opposition, until one day human control will be lost.

Why would we do such a thing, which causes our human downfall, when we each have a will to survive? Well, we certainly have a short-term will to survive, but whether it be military, economic, commercial or even purely scientific, we also have a will to progress, to get better, to get one step ahead of our neighbour or generally to improve ourselves. Franklin D. Roosevelt[69] said, 'The trend of civilisation itself is forever upward.' The short-term will to survive, which drives humans onwards and forwards, is in reality also the will to evolve and change and improve. It is the same will which in time will lead to our evolutionary descendants, our betters. Essentially we will evolve ourselves out of our bodies. Time will move on, possibly with machines at the helm, and humans will have filled their ecological and evolutionary niche. Humans will not even get a mention in the history books, as machines will probably use history CDs and not anything as primitive as books.

As far as the military situation is concerned, opposing sides will continue to improve their machines by giving them more and more intelligence and control. Networked computers will control missiles and combat aircraft, which will oppose each other in battle. All of these will have the fire power to end the lives of many humans. Prime targets for the machines will not be civilians with no controlling interests, but will rather be the opposition's machinery and even the opposition's humans who still have some degree of control. But the competition is there. Each side *must* stay a little ahead, each side must include more and more intelligence in order to do so, simply because machine intelligence is better than human intelligence where it matters. It is much faster and much more accurate. If the opposition is one step ahead and you know that in order to compete you must give your machines a little more intelligence, even if this means you may lose control of them, then what do you do? Do you remain subservient to the opposition, or do you take a risk and hope that it will turn out all right? After all, it has been all right for the human race as a whole up till now.

This is very much like blindly walking off the edge of a cliff. Before taking each step forward you say, 'The last step I took was OK, so I will take another one.' You remain safe until you reach the edge of the cliff, but then it is too late. But when introducing more intelligence into machines, we do not know where the edge of the cliff is, because our measures of intelligence are so fuzzy and need to cover such a broad area. So more intelligence will be introduced into machines. They will become more and more intelligent, until human and machine intelligence are comparable in performance. But that is one step too far.

In the present-day military set-up, human soldiers and pilots are used merely as extensions to machines,[70] simply to make up for a machine's inadequacy. A weapon system needs a disciplined human only to fill in its weak points in aiming and firing. As soon as the system is able to perform these tasks better than a human, the human role becomes obsolete. The behaviour, social performance and working task of some soldiers is merely to optimise the performance of a weapon system. The soldier is required to act automatically in an unintelligent, robotlike way, with no self-values or beliefs.[70] Indeed, this is the requirement until a machine can perform the same task in a better way.

The military state of affairs is a more obvious and recognisable structure in terms of its operation than is ordinary domestic life. In the military case commands are networked out to human soldiers and/or machine soldiers, and the humans, just like the machines, are merely components in a system,[71] controlled and driven from within the network itself. This is possibly centred on one decision-making mechanism, a president or a computer, or with decision-making spread around the network. If orders or instructions are passed through the network to the soldiers, whether they be human or machine, their role is merely to carry out those instructions exactly. The only intelligence they require is that necessary to carry out orders, nothing more and nothing less. They must certainly not question the order. Their lives are, if necessary, thrown away in order to achieve an overall goal or target decided upon within the network. A missile is sent off to its certain destruction, and human soldiers or pilotless combat aircraft are sent off to their possible destruction, for the overall good of the home team, and the overall bad of the away team. Clearly the direction in which we are heading is for human soldiers to continue to be phased out and machines, with some intelligence, to be introduced in their place. After all, this will save home-team human lives! So we will have a military system with a network controlling pilotless aircraft, autonomous tanks and robot machine soldiers, with all the advantages that holds in comparison with human components.

We can also look at the question of who will control this network of machine warfare. At present it is humans, aided by machines, in the form of computers and communication systems, but we are marching rapidly to the cliff edge. The drive is on to give intelligent computer machinery more of a role in the decision-making, until before very long it will be given the task of making major decisions as well. This must be so because the home team needs to stay one step ahead of the opposition. So we will have intelligent networked computer systems controlling machine armies and aircraft. They will impose what has been called[72] 'a grid of control' on the planet.

We face a future with missiles not simply controlled by computers in terms of targeting, but also with the decision to fire being under 'machine control'. We also face a future with new, deadlier machines: military machines which cannot be reasoned or bargained with and which also show little regard for human life. Indeed why should they? Their values will be machine values. Rights and wrongs will be in terms of how they apply to machines, not humans. Survival will be in terms of the machine network as a whole and soldier machines, just like humans, will be expendable.

To infiltrate human defensive positions, in which yesterday's technology and subservient machines are used, new purpose-designed machines will appear. It is a small step from the missiles and pilotless aircraft of today to conceive of kamikaze robot aircraft, filled with nuclear explosives that use highly precise targeting to ensure they maximise their impact on the target. Indeed, in the missiles of today we probably have them already. And how will humans stop them? Yesterday's technology will not be good enough. Kamikaze robot autonomous vehicles can position themselves exactly where required and detonate immediately, when some specific event occurs or at a preset time. Again, this is not far removed from present-day mine and bomb warfare; the only difference is the controlling interest.

Thousands of small hexapod or wheeled bomb robots could easily infiltrate a city, scattering themselves to all important locations, detonating at a pre-specified strategic time. Micro-machine bomb robots could even be placed inside human bodies by means of food or drugs, the person concerned becoming a walking time bomb, living and behaving normally, unaware that at a preset time or following a preset event they will simply explode from within. The range of size and ability in military machines, from micro-machines to rocket missiles, opens up an enormous spectrum of possibilities that would simply be impossible for humans to defend against, even with the help of other machines subservient to humans.

But what about human destruction? What about a repeat of the mass killings seen at the end of the Second World War in Hiroshima and Nagasaki? Is that likely to happen again? The bombs dropped in Japan were both equivalent to igniting 20,000 tonnes of TNT.[73] Even the old technology, and now phased-out, Titan II missiles each contained a reported 9,000,000 tonne warhead, thereby being 450 times more destructive. In terms of size of bombs and impact effect there is no question, therefore, that, far from a repeat, a present-day exercise could be much worse. So we return to the same conclusion that our future is dependent on who or what is in control.

When, as is indicated in this book as a likely scenario, machines do gain control, then can we expect them to use the powerful weapons and missiles described against humans? Suppose, in a country the size of Japan, there were only chimpanzee inhabitants, numbering several millions. Suppose they were using the country as a base to attack and kill humans in large numbers, even to the extent of threatening human civilisation itself. Would we humans then hesitate to seek and destroy them, possibly even blowing the country and all its chimpanzee inhabitants to pieces with extremely powerful nuclear weapons, especially if we humans were not seriously affected and the threat was dramatically reduced or even removed? I believe the answer is clear: humans would use nuclear weapons against such chimpanzees if necessary. It is also worth remembering that less powerful, but nevertheless devastating, nuclear weapons have already been used to end the lives of many humans, never mind chimpanzees. So certainly we can expect machines, if they become more intelligent and gain control of military machines, to use powerful nuclear weapons against us humans, especially if the machines themselves are not seriously affected. Unless, that is, we immediately give up the fight and remain subservient to them, becoming their slaves.

It may be, however, that dominant machines would choose other means of human destruction, such as gassing or biological weapons. Their methods would most likely have little or no effect whatsoever on the machines themselves. On the assumption that animals other than humans would not take up a fighting stance against the machines, we humans would be the main target, although the machines might also come up against opposition from insects at the micro-machine level. It is difficult to imagine, though, what fire power a hive of bees could offer against machines.

It is likely, however, that many insects and animals could be little affected by dominant machines, in comparison with humans. Many might be killed as a by-product of gassings or bombings of humans. However, if they were in areas of no human resistance, then it may be they would be left alone, unless the machines expanded on the human passions of biological research and zoos, in which humans would probably be placed in a section with the other apes. It is also difficult to know whether or not machines would require leisure time, perhaps to follow such pastimes as human-hunting, or even using humans in a bull ring, gladiatorial style.

In Chapter 4 we looked at what robot machines can do now, at how in some ways machines are already much better than humans at some things, and yet still have difficulties in other areas. Copying the workings of a human hand was seen to be a particular aspect in which there is still a gap. Indeed, machine performance may never be directly comparable with that of humans in this respect.

In Chapter 5, meanwhile, we looked into human intelligence. It was felt unlikely that a machine brain would ever be able to operate in exactly the same way as a human brain. Also it was concluded that machine brains might never be conscious in exactly the same way that human brains are. Combining the findings of chapters 4 and 5, therefore, it would appear that we are very unlikely *ever* to get machines which are exact copies of humans, both physically and mentally.

What overall conclusions can we draw from this? Can we say, as some others appear to, that because machines are unlikely to become exact copies of humans, we do not need to worry, we do not have a problem? Somehow it has also been concluded that because of this, machines will always remain subservient to humans. But can we really say this? Dinosaurs could probably have come up with the same conclusions about humans or the apes, if we had existed during their time. Many, many years ago they could have concluded that because the apes, including humans, would most likely never look physically like dinosaurs, never have brains which worked in exactly the same way and were not conscious like dinosaurs, so the dinosaur position as the dominant life form on earth was secure. Clearly just as dinosaurs would, in hindsight, have been wrong to draw such a conclusion, so too would we humans be wrong now. If the human race should have learned anything from its experience, it is that something better than us will come along on the Earth, at some time in the future, and that that something will, in all probability, be far more intelligent than we are.

As far as machines are concerned, they do not suffer from the vast array of physical limitations which humans must endure. They are not restricted by their size or a biological framework to anywhere near the same extent. They do not rely on a delicate balance of gases or temperatures for their survival. If taken only a relatively short distance from the Earth, a human life perishes unless it has the help of a machine. The present physical capabilities and future possibilities of machines allow a much broader range of life under a wider spectrum of habitats.

Machines also do not and will not necessarily suffer from the enormous mental restrictions that humans have. A human brain is of limited size. When taken out of a human body, according to our present medical knowledge, it is no longer operative. To continue functioning, it depends wholly on the physical body in which it resides. Humans cannot directly add to their brain's power, readily reprogram certain elements or channel their brain's operation in different ways. In essence, humans have little or no control over the operation of their brains.

Machine intelligence, meanwhile, is apparently limitless, something difficult for humans to conceive. Machines can simply go on getting more and more intelligent. It is quite possible that in the future machine intelligence could so far surpass human intelligence as to make the latter appear relatively insignificant. And yet it is this very human intelligence that has designed and developed machine intelligence in the first place. It is pushing the frontiers back continually and will, if progress continues along its present lines, realise machine intelligence which is at first roughly equivalent but which later surpasses the human level.

Perhaps just as in P.G. Wodehouse's stories,[74] in which the intelligent Jeeves is happy to work for the less clever Bertie Wooster, machines which are, overall, a little more intelligent than humans may still be prepared to be subservient to humans. It may be that their physical appearance or situation restricts them for a while. However, once machines have an intelligence which is at least as good as that of humans, then it is difficult to imagine that they would wish to remain subservient for long. Bertie Wooster had money and a position in human society which gave Jeeves reasons for his subservience. But what attributes do humans have that machines might like the look of? If indeed we have anything, it would not be long before the more intelligent machines found alternatives.

But machine intelligence is extremely difficult to pin down, categorise and measure, and will probably be even more so in the future. If quantifying human intelligence is not an easy task, then trying to do the same with machine intelligence, in a sensible and useful way, is probably futile. At least with human intelligence we can inspect one brain situated in one place at one time, and although we have considerable problems in detaching a body's physical attributes and capabilities from those of the brain which lies within, we can judge the body as a whole and try to minimise the effects of physical differences.

With machine intelligence it is relatively simple to inspect a stand-alone machine, such as a computer workstation operating in isolation or a Seven Dwarf robot, to apply tests such as the Turing test and to make direct comparisons between the intelligence witnessed and that experienced in humans or animals. However, in these last two chapters we have been looking more at networks, probably incorporating a variety of computational machines at different points, possibly fulfilling very different functions. At the same time, controlled by the network are robot machines which have some physical presence in the outside world and which may themselves, in stand-alone mode, be of relatively limited intelligence but have extremely powerful physical attributes.

We know that even the least intelligent humans, if they follow the instructions given to them by a much more intelligent person, can, in a Turing-like way, appear to be intelligent and can be extremely successful in what they do. It is the combined intelligence that is really important and not simply that of the individual. As previously highlighted, the same is true with a hive of bees. In both the human and the bee cases, though, what we arrive at is a combination of brains of a similar size, shape and function, and not very dissimilar performance. There is no direct, permanent physical link between the brains. It is essentially a case of 'Two heads are better than one.'

With machines the situation is, however, very different. Although it is possible for the computational machines connected into the network to be identical, this is not always the case. After all, if one machine is already fulfilling a role, it may not need to be coupled with another machine fulfilling exactly the same role, unless the same role is needed in a number of different places. We know from the basic concepts of mass production that it is better for each separate element in an overall scheme to be dealt with by a different individual. In this case, for efficiency in the network, this implies that it is best for different specialist machines to each carry out different roles required in the network as a whole.

So we can have different computational machines connected into the network, some carrying out identical roles whilst others carry out their own separately identifiable and different tasks. There will certainly be communication between the machines, although they can be not only distributed around the network but also located in completely different places. Also connected to the network we have robot machines, carrying out physical tasks required by the network as a whole and each one itself having a certain amount of intelligence which contributes to the network total. Of course there is nothing to stop one network being connected or linked to another, perhaps of a completely different form, perhaps complementary and perhaps similar.

It is quite possible, therefore, for us to view an intelligent machine network as a multi-neuron brain, with different areas of the brain dealing principally with their own tasks but with the ability to help out other areas. Communication between the areas is by very high-speed, possibly optical, links which can pass multiple signals at a time. The machine network is, therefore, like one big brain, spread over a wide space, but with different areas being able to communicate with each other much more rapidly than is the case with a human brain.

The question is, what is the machine intelligence that we are trying to measure? We can make an attempt at the intelligence of individual machines and even certain elements, but surely it is the network as a whole, the complete system, which is being viewed, and the intelligence of which is having an impact on the outside world. With machines it is really the total intelligence of the network that is important, as opposed to the intelligence of any individual machine in the network. This is akin to the total intelligence of a beehive being important rather than the intelligence of an individual bee.

With machine intelligence, we are looking at something which is potentially extremely powerful, which is potentially much bigger and more dense, not just in comparison with a beehive but also when compared with a group of humans. We are looking at something which is potentially much better connected, with a far superior and efficient communication network, and which houses a range of considerable advantages as far as individual characteristics of intelligence are concerned. In machine intelligence, we are facing something which is potentially far more intelligent than any known life form, including humans.

A major problem exists in that we humans do not really know, and certainly cannot agree between ourselves, what intelligence actually is, although we can witness some of its effects. We do not really know what it means to be conscious, although we can look at some indications. We do not really know what our purpose is on Earth, why we are here. So how can we know, in reality, when a networked machine intelligence is itself in control of a total intelligence which is superior to that of networked humans? How can we say that machines do not have a self-willed purpose in life when we do not know what our own purpose is, if we have one? Indeed, we may almost be at the point where machine intelligence is superior, but we simply will not know when it actually happens, because we cannot measure it.

The US Strategic Defense Initiative (SDI) of the mid-1980s, also referred to as Star Wars by the media, picked up on the networked intelligence and information aim[67] to try and achieve a defensive shield to protect the USA from all but minor damage as a result of a Soviet nuclear weapons attack. The aim was physically or electronically to destroy the nuclear weapons while they were still distant from the US, possibly in space. The shield would largely operate autonomously by means of decisions made by intelligent computing machines in order to prioritise and destroy detected threats rapidly, using advanced and futuristic weapons. This was seen, and indeed is still seen,[67] merely as part of the ongoing US military strategy and not as a distinct break from tradition.

It is interesting that this shift towards military battles in space has led some to believe that it is space that will be inherited by robot machines, with robots working and living entirely in this environment. There is much to be said for this. Humans could actually put the robot machines there, both for Star Wars purposes and as general worker robots, repairing space ships, manning (or should it be roboting) space stations and operating on distant planets. Humans themselves are well tuned solely to Earth operation. We can only operate in space with the help of machines of one kind or another. Essentially we are not well suited to space. Machines, on the other hand, can be designed and evolved to operate effectively in space, where they can settle and 'make themselves at home'. Without question, machines can be better suited to space than are humans.

But as far as the Earth is concerned, whether or not robot machines inherit space is a relative side issue, although having inherited space, the machines could later send a task force to Earth. It is easy to look out into the voids of space, where humans have barely dared to venture, and see a time in which robot machines are the dominant life form, at least in our own galaxy. It does not present humans with an immediate problem if robot machines dominate space, as we humans are not much in evidence there due to the hostile environment. But these robot machines we are sending out into space, with a backpack and a one-way ticket, have evolved from humans and will have, in one way or another, elements of human ways of looking at things. Why would they, therefore, be content to remain solely in space?

The theory of robot machines being dominant in space seems to imply that machines will, therefore, not dominate Earth. For some reason humans will say, 'OK, we have had enough of intelligent machines. Let's send them all out into space!' In this way the future of the human race is assured. Clearly this is not a realistic scenario. Whether or not robot machines dominate space does not mean that humans will suddenly forget about progress and return to a mediaeval way of life. No, we on Earth have become a society too reliant on intelligent machines to retreat to a more placid lifestyle. Our way of life, particularly in the western world, is far too dependent on machines for us to reverse the situation and do without things that we now take for granted. Whether or not machines dominate space does not affect the basic drive of humans on Earth to improve their lot, in one sense by enabling machines to do more and more things, allowing them to become progressively more intelligent.

With the robots at Reading University we have witnessed machines that learn, communicate with each other, teach each other and behave in different ways. We can see that humans are not particularly special, not particularly wonderful, not the ultimate intelligent beings. Rather, we are just a step on the road of evolution, biological beings with a relatively small range of operation, and in an extremely vulnerable position. We cannot hide behind a philosophical cloak whilst walking towards the cliff edge and say we are safe so far, we are better and more intelligent than anything else on Earth thus far, and that means we always will be. Our philosophy is not a human defensive shield, a virtual SDI, but rather could aid our downfall. We certainly cannot say that because everything has been in order until now, therefore we do not have to worry about our future.

And what is that future? What might it hold? There appears to be absolutely nothing to stop machines becoming more intelligent, particularly when we look towards an intelligent machine network. There is no proof, no evidence, no physical or biological pointers that indicate that machine intelligence cannot surpass that of humans. Indeed, it is ridiculous to think so. All the signs are that we will rapidly become merely an insignificant historical dot.

It looks unlikely that we will see humanoid robots which completely replicate humans. There are considerable technical difficulties with this, and little or no driving force. But we cannot conclude that, because machines are unlikely to be approximately equivalent to humans, they will always be subservient to us. In fact, the converse is true: it is *because* they are different, *because* they have distinct advantages, many of which we know about already, that machines can be better than we are. In this way they can dominate us physically through their superior intelligence.

The human race, as we know it, is very likely in its end game; our period of dominance on Earth is about to be terminated. We can try and reason and bargain with the machines which take over, but why should they listen when they are far more intelligent than we are? All we should expect is that we humans are treated by the machines in the same way that we now treat other animals, as slave workers, energy producers or curiosities in zoos. We must obey their wishes and live only to serve all our lives, what there is of them, under the control of machines.

As *the* human race, we are delicately positioned. We have the technology, we have the ability, I believe, to create machines that will not only be as intelligent as humans but that will go on to be far more intelligent still. This will spell the end of the human race as we know it. Is that what we want? Should we not at least have an international body monitoring and even controlling what goes on?

When the first nuclear bombs were dropped on Japan, killing thousands of people, we took stock of our actions and realised the threat that such weapons posed to our existence. Despite the results achieved by the Hiroshima and Nagasaki bombs, even deadlier nuclear bombs have been built, much more powerful, much more accurate and much more intelligent. But with nuclear weapons we saw what they could do and we gave ourselves another chance.

With intelligent machines we will not get a second chance. Once the first powerful machine, with an intelligence similar to that of a human, is switched on, we will most likely not get the opportunity to switch it back off again. We will have started a time bomb ticking on the human race, and we will be unable to switch it off.

Bibliography

1. Palfreman, J. and Swade, D. (1991): *The Dream Machine*. BBC Books.
2. Zeman, A. (1995): 'Hope for a Life Free of Seizures'. *The Times*, 31 October, p. 16.
3. Lettvin, J.Y., Maturana, H.R., McCulloch, W.S. and Pitts, W.H. (1959): 'What the Frog's Eye tells the Frog's Brain'. *Proc. IRE*, pp. 1940–51.
4. Eysenck, H.J. (1966): *Check Your Own IQ*. Penguin.
5. Shaw, G.B. (1996): *Pygmalion*. New Longman Literature.
6. Čapek, K. (1940): *Rossum's Universal Robots*. Fr. Borovy, Prague.
7. Smart, J.J.C. (1959): 'Professor Ziff on Robots.' *Analysis*, Vol. XIX, No. 5.
8. Dawkins, R. (1976): *The Selfish Gene*. Oxford University Press.
9. Prabhupada, A.C.B.S. (1994): *Beyond Birth and Death*. Bhaktivedanta Book Trust, Bombay.
10. McKenna, A. (1995) 'Our Bizarre Double Lives.' *What's on TV*, October 1995, IPC Magazines,
11. Hirose, S. (1993): *Biologically Inspired Robots*. Oxford University Press.
12. Scott-Stakes, H. (1982): 'Japan's Love Affair with the Robot.' *New York Times Magazine*, 10 January 1982, p.26.
13. Kendall, D. (1995): 'The Robot is in the Post'. *Computing & Control Engineering Journal*, Vol. 6, No. 4, pp. 177–81.
14. Jacobsen, S.C., Wood, J.E., Knutti, D.F., Biggers, K.B. and Iversen, E.K. (1985): 'The Version 1 Utah/MIT Dextrous Hand.' In *Robotics Research: The Second International Symposium*, H. Hanafusa and H. Inoue (eds.), MIT Press, pp. 301–8.
15. Bekey, G.A., Tomovic, R. and Zeljkovic, I. (1990): 'Control Architecture for the Belgrade/USC Hand.' In *Dextrous Robot Hands*, S.T. Venkataraman and T. Iberall, (eds.) Springer-Verlag, pp. 136–49.

16. Salisbury, J.K. (1982): *Kinematic and Force Analysis of Articulated Hands*. Ph.D. Dissertation, Department of Mechanical Engineering, Stanford University.
17. Al-Gallaf, E., Allen, A.J. and Warwick, K. (1993): 'Dextrous Hands: Issues Relating to a Four-Fingered Articulated Hand.' *Mechatronics*, Vol. 3, pp. 329–42.
18. Brady, J.M. and Paul, R.P. (1984): *Int. Journal of Robotics Research*, Vol. 3. No. 2, Special Issue on Walking Machines, Guest Editor M.H. Raibert. Note: Several later issues also contain relevant articles.
19. Raibert, M.H. (1986): *Legged Robots that Balance*. MIT Press, Cambridge, Mass.
20. Archer, N.J. (1993): *Control Aspects of Bipedal Walking*. Ph.D. Thesis, Reading University.
21. Bodkin, W. (1995): 'Joe's Robot Winner'. *Financial Mail on Sunday*, 10 December 1995, p.5.
22. Trevelyan, J.P. (1989): 'Sensing and Control for Sheep-Shearing Robots'. *IEEE Transactions on Robotics and Automation*, Vol. 5, pp. 716–27.
23. Kroczynsky, P. and Wade, B. (1987): 'The Skywasher: a Building Washing Robot'. *Robots 11: Seventeenth International Symposium on Industrial Robots*, Chicago, Society of Manufacturing Engineers, April 1987, pp. 11–19.
24. Engelberger, J.F. (1989): *Robotics in Service*. Kogan Page, London.
25. Moravec, H. (1988): *Mind Children: The Future of Robot and Human Intelligence*. Harvard University Press.
26. Malcom, N. (1970): 'Scientific Materialism and the Identity Theory'. In C. Borst (ed.), *The Mind/Brain Identity Theory*, Macmillan, London.
27. Minsky, M. (ed.) (1968): *Semantic Information Processing*. MIT Press, Cambridge, Mass.
28. Kelly, J. (1993): *Artificial Intelligence: A Modern Myth*. Ellis Horwood.
29. Macmillan Encyclopedia (1994): definition of 'Computer'. Macmillan.
30. Chandor, A. (1977): *A Dictionary of Computers*. Penguin.
31. Gloess, P.Y. (1981): *Understanding Artificial Intelligence*. Alfred Publishing, California.
32. Barr, A. and Feigenbaum, E.A. (eds.) (1981): *The Handbook of Artificial Intelligence*. Vol. 1, Pitman, London.
33. Boden, M.A. (1987): *Artificial Intelligence and Natural Man* (2nd edn.). MIT Press, Cambridge, Mass.
34. Penrose, R. (1994): *Shadows of the Mind*. Oxford University Press.

35. Kant, I. (1978): *Critique of Pure Reason*. Translated by Norman Kemp Smith, Garland Publishers, New York.
36. Bundy, A. (1995): Letter to *New Scientist*, 18 September 1995.
37. Dreyfus, H.J. and Dreyfus, S.E. (1990): 'Making a Mind Versus Modelling the Brain: Artificial Intelligence Back at a Branch Point'. In *The Philosophy of Artificial Intelligence*, M.A. Boden (ed.), Oxford University Press.
38. Dennett, D.C. (1990): 'Cognitive Wheels: The Frame Problem of AI'. In *The Philosophy of Artificial Intelligence*, M.A. Boden (ed.), Oxford University Press.
39. Dennett, D. (1991): *Consciousness Explained*. Penguin.
40. Edelman, G.M. (1989): *The Remembered Present: A Biological Theory of Consciousness*. Basic Books, New York.
41. Searle, J.R. (1981): 'Minds, Brains and Programs'. In *The Mind's I*, D.R. Hofstadter and D.C. Dennett (eds.), Basic Books Inc., Penguin.
42. Deutsch, D. (1989): 'Quantum Computational Networks'. *Proc. Royal Society*, London, A425, pp. 73–90.
43. Rosenblatt, F. (1962): *Principles of Neurodynamics: Perceptrons and the Theory of Brain Mechanisms*. Spartan, Washington DC.
44. Orwell, G. (1948): *1984*. Penguin.
45. Seymore, J. and Norwood, D. (1993): 'A Game for Life'. *New Scientist*, Vol. 139, No. 1889, pp. 23–6.
46. Holmes, B. (1996): 'The Creativity Machine'. *New Scientist*, 20 January 1996, pp. 22–6.
47. Monod, J. (1972): *Chance and Necessity*. Collins, London.
48. Simons, G. (1992): *Robots: The Quest for Living Machines*. Cassell.
49. Ordish, R. (1992): *The Jim'll Fix It Story*. Hodder & Stoughton.
50. Graham, D. (1985): 'Pattern Control of Walking Insects'. *Advances in Insect Physiology*, pp. 31–140.
51. Raibert, M. and Sutherland, I.E. (1983): 'Machines that Walk'. *Scientific American*, Vol. 248, pp. 32–41.
52. Hildebrand, J. (1976): 'Analysis of Tetrapod Gaits'. *Neural Control of Locomotion*, Plenum Press, pp. 203–36.
53. Brooks, R. (1986): 'A Robust Layered Control System for a Mobile Robot'. *IEEE Trans. on Robotics & Automation*, Vol. RA-2, pp. 14–23.
54. Batchelor, B.G. and Waltz, F.M. (1993): *Interactive Image Processing for Machine Vision*. Springer-Verlag.
55. Hopfield, J.T. (1982): 'Neural Networks and Physical Systems with

Emergent Collective Computational Abilities'. *Proc. National Acad. Science, USA*, Vol. 79, pp. 2554–8.

56. Bishop, J.M., Minchinton, P.R. and Mitchell, R.J. (1991): 'Real Time Invariant Grey Level Image Processing Using Digital Neural Networks'. *Proc. IMechE Eurodirect '91*, London, pp. 187–8.

57. Berkeley, G. (1979): *Three Dialogues between Hylas and Philonous*. R.M. Adams (ed.), Hackett, Indianapolis.

58. Asimov, I. (1986): *I Robot*. Grafton Books.

59. de Garis, H. (1995): 'CAM-BRAIN'. *Proc. Int. Conference on Artificial Neural Nets and Genetic Algorithms*, Ales, France, Springer-Verlag, pp. 84–7.

60. Gibson, W. (1995): *Mona Lisa Overdrive*. Voyager.

61. Nietzsche, F. (1961): *Thus Spoke Zarathustra*. Penguin Classics.

62. Darwin, C. (1995): *Origin of Species by Means of Natural Selection*. Grammercy Books.

63. Mullins, J. (1996): 'One False Step'. *New Scientist*, No. 2028, 4 May 1996, pp. 32–7.

64. King, D. (1996): talking on BBC TV about Frank Bruno's chances of beating Mike Tyson for the Heavyweight Championship of the World.

65. Booth, M. (1995): personal communication.

66. *Air International* (1996), Vol. 50, No. 5, Key Publishing.

67. Bonds, R. (1987): *The Modern US War Machine*. Salamander.

68. Munro, R. and Chant, C. (1995): *Jane's Combat Aircraft*. HarperCollins.

69. Roosevelt, F.D. (1944): presidential address.

70. Radine, L. (1977): *The Taming of the Troops: Social Control in the US Army*. Greenwood Press, Westport CT.

71. Levidow, L. and Robins, K. (eds.) (1989): *Cyborg Worlds: The Military Information Society*. Free Association Books, London.

72. Haraway, D. (1985): 'A Manifesto for Cyborgs: Science, Technology and Socialist Feminism in the 1980s'. *Socialist Review*, Vol. 80, pp. 65–107.

73. Macksey, K. and Woodhouse, W. (1993): *The Penguin Encyclopedia of Modern Warfare*. Penguin.

74. Wodehouse, P.G. (1965): *Jeeves in the Offing*. Penguin.

Further Reading

The following is a list of suggested texts for those wishing to go deeper into some of the topics discussed in this book.

Artificial Intelligence

Boden, M.A. (ed.) (1990): *The Philosophy of Artificial Intelligence*. Oxford University Press, especially the chapters by Turing, Searle, Sloman and Margaret Boden herself.

Neural Networks

Aleksander, I. and Morton, H. (1993): *Neurons and Symbols*. Chapman and Hall.

Chichocki, A and Unbehauen, R. (1993): *Neural Networks for Optimisation and Signal Processing*. John Wiley and Sons.

Irwin, G.W., Warwick, K. and Hunt, K.J. (eds.) (1995): *Neural Network Applications in Control*. Peter Peregrinus Ltd.

Consciousness and the Mind

Crick, F. and Koch, C. (1992): 'The Problem of Consciousness'. *Scientific American*, Vol. 267, p. 110.

Searle, J.R. (1992): *The Rediscovery of the Mind*. MIT Press, Cambridge, Mass.

Control

Gupta, M.M. and Sinha, N.K. (eds.) (1996): *Intelligent Control Systems*. IEEE Press.

Warwick, K. (1996): *An Introduction to Control Systems*. World Scientific Press.

Robots and Artificial Life

Gaussier, P. (ed.) (1995): *Journal on Robotics and Autonomous Systems*. Special issue on Moving the Frontiers between Robotics and Biology, Vol. 16, December 1995, especially the papers by Ferrell and Matovic.

Kevin Warwick is Professor of Cybernetics at the University of Reading. He carries out research in artificial intelligence, control and robotics. His favourite topic is pushing back the frontiers of machine intelligence. Kevin was born in Coventry in 1954 and went to school in Rugby before taking up a non-academic career by joining British Telecom, with whom he spent the next six years. At twenty-two he took his first degree at Aston University followed by a PhD and a research post at Imperial College, London. He subsequently held positions at Oxford, Newcastle and Warwick Universities before taking up the Professorship at Reading in 1988. Kevin, who has been awarded higher Doctorates both by Imperial College and the Czech Academy of Sciences, Prague, has been described as Britain's leading prophet of the robot age. He has two children and lives with his Czech wife in Reading.

The University of Illinois Press
is a founding member of the
Association of American University Presses.

———————————————————

University of Illinois Press
1325 South Oak Street
Champaign, IL 61820-6903
www.press.uillinois.edu